An Introduction to Applied Optimal Control

This is Volume 159 in
MATHEMATICS IN SCIENCE AND ENGINEERING
A Series of Monographs and Textbooks
Edited by RICHARD BELLMAN, *University of Southern California*

The complete listing of books in this series is available from the Publisher upon request.

An Introduction to Applied Optimal Control

Greg Knowles

Department of Mathematics
Carnegie-Mellon University
Pittsburgh, Pennsylvania

1981

ACADEMIC PRESS

A Subsidiary of Harcourt Brace Jovanovich, Publishers

New York London Toronto Sydney San Francisco

ACADEMIC PRESS, INC.
111 Fifth Avenue, New York, New York 10003

United Kingdom Edition published by
ACADEMIC PRESS, INC. (LONDON) LTD.
24/28 Oval Road, London NW1 7DX

Library of Congress Cataloging in Publication Data

Knowles, Greg.
An introduction to applied optimal control.

(Mathematics in science and engineering)
Includes bibliographies and index.
1. Control theory. 2. Mathematical optimization.
I. Title. II. Series.
QA402.3.K56 629.8'312 81-7989
ISBN 0-12-416960-0 AACR2

PRINTED IN THE UNITED STATES OF AMERICA

81 82 83 84 9 8 7 6 5 4 3 2 1

Contents

Preface

This book began as the notes for a one-semester course at Carnegie-Mellon University. The aim of the course was to give an introduction to deterministic optimal control theory to senior level undergraduates and first-year graduate students from the mathematics, engineering, and business schools. The only prerequisite for the course was a junior level course in ordinary differential equations. Accordingly, the backgrounds of the students were widely dissimilar, and the common denominator was their interest in the applications of optimal control theory. In fact, one of the most popular aspects of the course was that students were able to study problems from areas that they would not normally cover in their respective syllabi.

This text differs from the standard ones in that we have not attempted to prove the maximum principle, since this was beyond the background and interest of most of the students in the course. Instead we have tried to show its strengths and limitations through examples.

In Chapter I we introduce the concept of optimal control by means of examples. In Chapter II necessary conditions for optimality for the linear time optimal control problem are derived geometrically, and illustrations are given. In Chapters III and IV we discuss the Pontryagin maximum principle, its relation to the calculus of variations, and its application to various problems in science, engineering, and business. Since the optimality conditions arising from the maximum principle can often be solved only numerically, numerical techniques are discussed in Chapter V. In Chapter VI the dynamic programming approach to the solution of optimal control problems and differential games is considered; in Chapter VII the controllability and observability of linear control systems are discussed, and in Chapter VIII the extension of the maximum principle to state-constrained control problems is given. Finally, for more advanced students with a background in functional analysis, we consider in Chapter IX several problems in the control of systems governed by partial differential equations. This could serve as an introduction to research in this area.

The support of my colleagues and students at Carnegie-Mellon University has been invaluable during this project; without it this text would almost certainly not have appeared.

Chapter I | Examples of Control Systems; the Control Problem

Example 1 Consider a mechanism, such as a crane or trolley, of mass m which moves along a horizontal track without friction. If $x(t)$ represents the position at time t, we assume the motion of the trolley is governed by the law

$$m\ddot{x}(t) = u(t), \qquad t > 0, \tag{1}$$

where $u(t)$ is an external controlling force that we apply to the trolley (see Fig. 1). Assume that the initial position and velocity of the trolley are given as $x(0) = x_0, \dot{x}(0) = y_0$, respectively. Then we wish to choose a function u (which is naturally enough called a control function) to bring the trolley to rest at the origin in minimum time. Physical restrictions will usually require that the controlling force be bounded in magnitude, i.e., that

$$|u(t)| \leq M. \tag{2}$$

For convenience, suppose that $m = M = 1$, and rewrite Eq. (1)

$$\dot{x}_1 = x_2, \qquad \dot{x}_2 = u(t),$$

where $x_1(t)$ and $x_2(t)$ are now the position and velocity of the body at time t. Equation (1) then becomes

$$\begin{bmatrix} \dot{x}_1 \\ \dot{x}_2 \end{bmatrix} = \begin{bmatrix} 0 & 1 \\ 0 & 0 \end{bmatrix} \begin{bmatrix} x_1 \\ x_2 \end{bmatrix} + \begin{bmatrix} 0 \\ 1 \end{bmatrix} u(t)$$

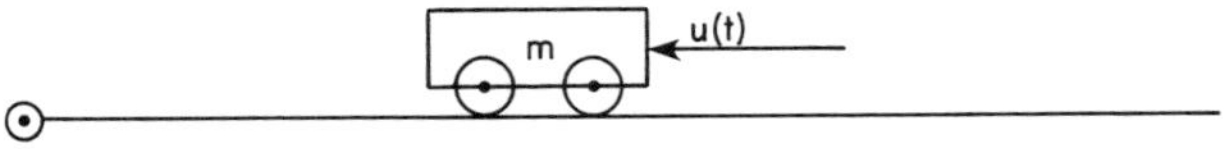

Fig. 1

or

$$\dot{\mathbf{x}}(t) = A\mathbf{x}(t) + \mathbf{b}u(t), \qquad \mathbf{x}(0) = \begin{bmatrix} x_0 \\ y_0 \end{bmatrix}, \tag{3}$$

where

$$A = \begin{bmatrix} 0 & 1 \\ 0 & 0 \end{bmatrix} \quad \text{and} \quad \mathbf{b} = \begin{bmatrix} 0 \\ 1 \end{bmatrix}, \quad \mathbf{x}(t) = \begin{bmatrix} x_1(t) \\ x_2(t) \end{bmatrix},$$

and the control problem is to find a function u, subject to (2), which brings the solution of (3), $\mathbf{x}(t)$, to $\left[\begin{smallmatrix} 0 \\ 0 \end{smallmatrix}\right]$ in minimum time t. Any control that steers us to $\left[\begin{smallmatrix} 0 \\ 0 \end{smallmatrix}\right]$ in minimum time is called an optimal control. Intuitively, we should expect the optimal control is first a period of maximum acceleration ($u = +1$), and then maximum braking ($u = -1$), or vice versa.

Example 2 (Bushaw [1]) A control surface on an aircraft is to be kept at rest at a fixed position. A wind gust displaces the surface from the desired position. We assume that if nothing were done, the control surface would behave as a damped harmonic oscillator. Thus if θ measures the deviation from the desired position, then the free motion of the surface satisfies the differential equation

$$\ddot{\theta} + a\dot{\theta} + \omega^2\theta = 0$$

with initial conditions $\theta(0) = \theta_0$ and $\dot{\theta}(0) = \theta'_0$. Here θ_0 is the displacement of the surface resulting from the wind gust and θ'_0 is the velocity imparted to the surface by the gust. On an aircraft the oscillation of the control surface cannot be permitted, and so we wish to design a servomechanism to apply a restoring torque and bring the surface back to rest in minimum time. The equation then becomes

$$\ddot{\theta}(t) + a\dot{\theta}(t) + \omega^2\theta(t) = u(t), \qquad \theta(0) = \theta_0, \qquad \dot{\theta}(0) = \theta'_0, \tag{4}$$

where $u(t)$ represents the restoring torque at time t. Again we must suppose that $|u(t)| \leq C$, where C is a constant, and by normalization can be taken as 1. The problem is then to find such a function u, so that the system will be brought to $\theta = 0$, $\dot{\theta} = 0$ in minimum time.

It is clear that if $\theta_0 > 0$ and $\theta'_0 > 0$, then the torque should be directed initially in the direction of negative θ and should have the largest possible magnitude. Thus $u(t) = -1$ initially. However, if $u(t) = -1$ is applied for too long a time, we shall overshoot the desired terminal condition $\theta = 0$, $\dot{\theta} = 0$. Therefore at some point there should be a torque reversal to $+1$ in order to brake the system.

The following questions occur:

(1) Is this strategy indeed optimal, and if so, when should the switch take place?

(2) Alternatively, is it better to remove the torque at some point, allow a small overshoot, and then apply $+1$?

(3) In this vein, we could ask whether a sequence of -1, $+1$, -1, $+1$, ... of n steps is the best, and if so, what is n and where do the switches occur?

Again we are led to controls that take on (only) values ± 1; such controls are called *bang-bang* controls.

Note that as before, setting $x_1 = \theta$ and $x_2 = \dot{\theta}$, we can write the system equation (4)

$$\dot{x}_1 = x_2, \qquad x_1(0) = \theta_0,$$

$$\dot{x}_2 = -ax_2 - \omega^2 x_1 + u, \qquad x_2(0) = \theta'_0,$$

$$\dot{\mathbf{x}} = A\mathbf{x} + \mathbf{b}u, \qquad \mathbf{x}(0) = \begin{bmatrix} \theta_0 \\ \theta'_0 \end{bmatrix},$$

where

$$A = \begin{bmatrix} 0 & 1 \\ -\omega^2 & -a \end{bmatrix}, \qquad \mathbf{b} = \begin{bmatrix} 0 \\ 1 \end{bmatrix},$$

and u is chosen with $|u(t)| \leq 1$ and to minimize

$$C(u) = \int_0^{t_1} 1\, dt,$$

where t_1 is any time for which $x_1(t_1) = 0$ and $x_2(t_1) = 0$.

Example 3 (Isaacs [3]) Let $x(t)$ be the amount of steel produced by a mill at time t. The amount produced at time t is to be allocated to one of two uses:

(1) production of consumer products;
(2) investment.

It is assumed that the steel allocated to investment is used to increase productive capacity—by using steel to produce new steel mills, transport facilities, or whatever. Let $u(t)$, where $0 \leq u(t) \leq 1$, denote the fraction of steel produced at time t that is *allocated* to *investment*. Then $1 - u(t)$ represents the fraction allocated to consumption. The assumption that the reinvested steel is used to increase the productive capacity could be written

$$\frac{dx}{dt} = ku(t)x(t), \qquad \text{where} \quad x(0) = C - \text{initial endowment},$$

where k is an appropriate constant (i.e., rate of increase in production is proportional to the amount allocated to investment).

The problem is to choose $u(t)$ so as to maximize the total consumption over some fixed period of time $T > 0$. That is, we are to maximize

$$\int_0^T (1 - u(t))x(t)\, dt.$$

For this problem, do we consume everything produced, or do we invest some at present to increase capacity now, so that we can produce more and hence consume more later? Do we follow a bang-bang procedure of first investing everything and then consuming everything?

Example 4 *Moon-Landing Problem* (Fleming and Rishel [2]) Consider the problem of a spacecraft attempting to make a soft landing on the moon using the minimum amount of fuel. For a simplified model, let m denote the mass, h the height, v the vertical velocity of the spacecraft above the moon, and u the thrust of the spacecraft's engine (m, h, v, and u are functions of time). Let M denote the mass of the spacecraft without fuel, h_0 the initial height, v_0 the initial velocity, F the initial amount of fuel, α the maximum thrust of the engine, k a constant, and g the gravity acceleration of the moon (considered constant). The equations of motion are

$$\begin{aligned} \dot{h} &= v, \\ \dot{v} &= -g + m^{-1}u, \\ \dot{m} &= -ku, \end{aligned}$$

and the control u is restricted so that $0 \le u(t) \le \alpha$. The end conditions are

$$\begin{aligned} h(0) &= h_0, \\ v(0) &= v_0, \\ m(0) - M - F &= 0, \\ h(t_1) &= 0, \\ v(t_1) &= 0, \end{aligned}$$

where t_1 is the time taken for touchdown.

With

$$x_1 = h, \qquad x_2 = v, \qquad x_3 = m,$$
$$\mathbf{x}(0) = (h_0, v_0, M + F)^{\mathrm{T}},$$
$$\mathbf{x}(t_1) = (0, 0, \text{anything})^{\mathrm{T}},$$

this problem becomes, in matrix form,

$$\dot{\mathbf{x}} = \begin{bmatrix} x_2 \\ -g + x_3^{-1}u \\ -ku \end{bmatrix} = \mathbf{f}(t, \mathbf{x}, u),$$

and we wish to choose u so that $0 \le u(t) \le \alpha$ and

$$-M - F - \int_0^{t_1} \dot{x}_3(\tau)\, d\tau \quad (= -m(t_1))$$

is a minimum.

However, $\dot{x}_3 = -ku$, so the above becomes

$$-M - F + k \int_0^{t_1} u(\tau)\, d\tau,$$

and this is minimized at the same time as

$$C(u) = \int_0^{t_1} u(\tau)\, d\tau.$$

Note that although these problems come from seemingly completely different areas of applied mathematics, they all fit into the following general pattern.

GENERAL FORM OF THE CONTROL PROBLEM

(1) The *state equation is*

$$\dot{x}_i = f_i(t, x_1, \ldots, x_n, u_1, \ldots, u_m), \qquad i = 1, \ldots, n,$$

or in vector form

$$\dot{\mathbf{x}} = \mathbf{f}(t, \mathbf{x}, \mathbf{u}),$$

where

$$\mathbf{x} = \begin{bmatrix} x_1 \\ \vdots \\ x_n \end{bmatrix} \quad \text{and} \quad \mathbf{u} = \begin{bmatrix} u_1 \\ \vdots \\ u_m \end{bmatrix}.$$

(2) The *initial point* is $\mathbf{x}(0) = \mathbf{x}_0 \in R^n$, and the *final point* that we wish to reach is $\mathbf{x}_1 \in R^n$. The final point $\mathbf{x}_1$ is often called the target (point), and may or may not be given.

(3) The class Δ of *admissible controls* is the set of all those control functions u allowed by the physical limitations on the problem. (In Examples 1 and 2 we had $\Delta = \{u : |u(t)| \leq 1\}$ and $m = 1$.) Usually we shall be given a compact, convex set $\Omega \subset R^m$ (the *restraint set*) and we shall take

$$\Delta = \{\mathbf{u} = (u_1, \ldots, u_m) : u_i \text{ piecewise continuous and } \mathbf{u}(t) \in \Omega\}.$$

(4) The *cost function* or *performance index* quantitatively compares the effectiveness of various controllers. This is usually of the form

$$C(\mathbf{u}) = \int_0^{t_1} f_0(t, \mathbf{x}(t), \mathbf{u}(t))\, dt,$$

where f_0 is a given continuous real-valued function, and the above integral is to be interpreted as: we take a control $\mathbf{u} \in \Delta$, solve the state equations to obtain the corresponding $\mathbf{x}$, calculate f_0 as a function of t, and perform the integration. If a target point is given (so called *fixed-end-point* problem), then t_1 must be such that $\mathbf{x}(t_1) = \mathbf{x}_1$. In particular, if $f_0 \equiv 1$, then $C(u) = t_1$, and we have the minimum-time problem. If a target point is not given (*free-end-point* problem), then t_1 will be a fixed given time, and the integration is performed over the fixed interval $[0, t_1]$.

The optimal control problem can now be formulated: Find an admissible control $\mathbf{u}^*$ that minimizes the cost function, i.e., for which $C(\mathbf{u}^*) \leq C(\mathbf{u})$ for all $\mathbf{u} \in \Delta$. Such controls $\mathbf{u}^*$ are called *optimal controls.*

We shall first investigate in depth in Chapter 2 the linear (i.e., state equations are linear in $\mathbf{x}$ and $\mathbf{u}$) time optimal control problem, deriving a necessary condition for optimality known as Pontryagin's maximum principle [4].

References

[1] D. Bushaw, Optimal discontinuous forcing terms, *in* "Contributions to the Theory of Non-linear Oscillations," pp. 29–52. Princeton Univ. Press, Princeton, New Jersey, 1958.

[2] W. Fleming and R. Rishel, "Deterministic and Stochastic Optimal Control." Springer-Verlag, Berlin and New York, 1975.

[3] R. Isaacs, "Differential Games." Wiley, New York, 1965.

[4] L. Pontryagin, V. Boltyanskii, R. Gramkrelidze, and E. Mischenko, "The Mathematical Theory of Optimal Processes." Wiley (Interscience), New York, 1962.

For extensive bibliographies on control theory, the reader should consult

E. B. Lee and L. Markus, "Foundations of Optimal Control Theory." Wiley, New York, 1967.

M. Athans and P. Falb, "Optimal Control: An Introduction to the Theory and Its Applications. McGraw-Hill, New York, 1965.

Chapter II The General Linear Time Optimal Problem

1. INTRODUCTION

Consider a control system described by the vector differential equation

$$\dot{\mathbf{x}}(t) = A(t)\mathbf{x}(t) + B(t)\mathbf{u}(t), \qquad \mathbf{x}(0) = \mathbf{x}_0 \in R^n, \tag{1}$$

where

$$\mathbf{x}(t) = \begin{bmatrix} x_1(t) \\ \vdots \\ x_n(t) \end{bmatrix}, \qquad \mathbf{u}(t) = \begin{bmatrix} u_1(t) \\ \vdots \\ u_m(t) \end{bmatrix},$$

$$A(t) = \begin{bmatrix} a_{11}(t) & \cdots & a_{1n}(t) \\ a_{21}(t) & \cdots & a_{2n}(t) \\ \vdots & & \vdots \\ a_{n1}(t) & \cdots & a_{nn}(t) \end{bmatrix}, \qquad B(t) = \begin{bmatrix} b_{11}(t) & \cdots & b_{1m}(t) \\ b_{21}(t) & \cdots & b_{2m}(t) \\ \vdots & & \vdots \\ b_{n1}(t) & \cdots & b_{nm}(t) \end{bmatrix},$$

and we assume the elements of $A(t)$, $B(t)$ are integrable functions over any finite interval of time.

The set of admissible controls will be

$$\Delta = \{\mathbf{u} = (u_1, \ldots, u_m)^{\mathrm{T}} : |u_i(t)| \leq 1,\ i = 1, \ldots, m\}. \tag{2}$$

A target point $\mathbf{x}_1 \in R^n$ is given, and the control problem is to minimize the time t_1 for which $\mathbf{x}(t_1) = \mathbf{x}_1$.

From the theory of ordinary differential equations, the solution of (1) can be written

$$\mathbf{x}(t:\mathbf{u}) = X(t)\mathbf{x}_0 + X(t)\int_0^t X^{-1}(\tau)B(\tau)\mathbf{u}(\tau)\,d\tau, \tag{3}$$

where $X(t)$ is the principal matrix solution of the homogeneous system $\dot{X}(t) = A(t)X(t)$, $X(0) = I$, the identity matrix [3, 7].

At a given time t we define the *attainable set* at time t to be just the set of all those points $\mathbf{x} \in R^n$ that we can reach in time t using all of our admissible controls, i.e.,

$$\mathscr{A}(t) = \{\mathbf{x}(t:\mathbf{u}) : \mathbf{u} \in \Delta\}.$$

A knowledge of the attainable set at time t will give us a complete description of the points we can hit in time up to t (see Fig. 1). Further we can reformulate our control problem:

Minimize the time t_1 *for which* $\mathbf{x}_1 \in \mathscr{A}(t_1)$.

Before we proceed to characterize time optimal controls, we should really at this point ask ourselves, Does an optimal control exist? Otherwise we lay ourselves open to the possibility of constructing something that may not exist. This point should be stressed, for in a control problem we are usually trying to force a physical system to behave according to our requirements, and there is generally no reason to expect that nature will be sympathetic.

For the linear time optimal control problem, it is shown in Appendix 1 that optimal controls exist if the target point $\mathbf{x}_1$ can be hit in some time. This last condition brings us to the subject known as *controllability*, which will be further discussed in Chapter VII.

From now on we shall *assume* that an optimal control $\mathbf{u}^*$ exists, where t^* is the minimum time, and we shall consider the problem of

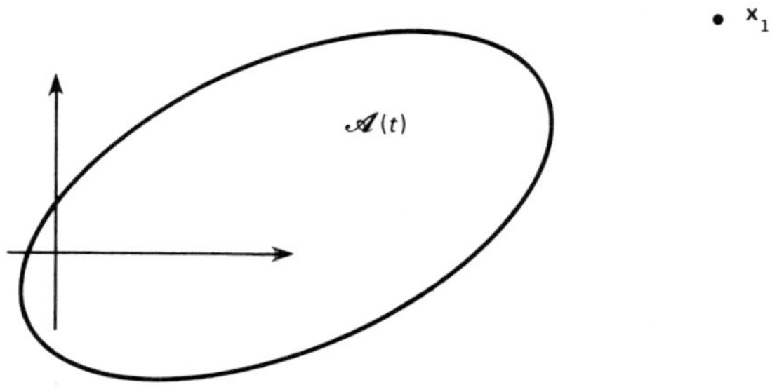

Fig. 1

characterizing $\mathbf{u}^*$. Further we shall consider only the case of one control variable, i.e., $m = 1$. B is then $n \times 1$. The general case is only notationally different.

Define $\mathbf{y}(t) = X^{-1}(t)B(t)$, i.e., $\mathbf{y}(t) = (y_1(t), \ldots, y_n(t))^{\mathrm{T}}$, and $\mathbf{y}_1 = X^{-1}(t^*)\mathbf{x}_1 - \mathbf{x}_0 \in R^n$, i.e., $\mathbf{y}_1 = \int_0^{t^*} \mathbf{y}(\tau)u^*(\tau)\,d\tau$. Set

$$\mathscr{R}(t) = \left\{\int_0^t \mathbf{y}(\tau)u(\tau)\,d\tau : u \in \Delta\right\}$$

$$= \left\{\left(\int_0^t y_1(\tau)u(\tau), \int_0^t y_2(\tau)u(\tau), \ldots, \int_0^t y_n(\tau)u(\tau)\,d\tau\right)^{\mathrm{T}} : u \in \Delta\right\}.$$

$\mathscr{R}(t)$ is called the *reachable set* in time t, and we show in Appendix 1 that $\mathscr{R}(t)$ is closed, bounded, and convex for all $t \geq 0$. In fact

$$\mathscr{A}(t) = X(t)[\mathbf{x}_0 + \mathscr{R}(t)] = \{X(t)[\mathbf{x}_0 + \mathbf{y}] : \mathbf{y} \in \mathscr{R}(t)\}$$

and

$$\mathscr{R}(t) = X^{-1}(t)\mathscr{A}(t) - \mathbf{x}_0.$$

Since $\mathbf{x}_1 \in \mathscr{A}(t^*)$, we have $\mathbf{y}_1 \in \mathscr{R}(t^*)$, and clearly t^* is the smallest time for which $\mathbf{y}_1 \in \mathscr{R}(t)$ (otherwise we would have $\mathbf{x}_1 \in \mathscr{A}(t)$ for $t < t^*$, a contradiction).

If we think of the control problem in terms of these sets $\mathscr{R}(t)$, $t > 0$, we know that at $t = 0$, $\mathscr{R}(t) = \{0\}$, and as time increases $\mathscr{R}(t)$ grows and eventually intersects the point $\mathbf{y}_1$. Our problem becomes, find the first time t^* at which $\mathscr{R}(t)$ intersects $\mathbf{y}_1$. We should expect, as is shown in Hermes [3] that the first contact occurs when $\mathscr{R}(t^*)$ just touches $\mathbf{y}_1$, or that $\mathbf{y}_1$ must belong to the boundary of $\mathscr{R}(t^*)$ (see Fig. 2).

The supporting hyperplane theorem (Appendix 1) then implies that there is a nontrivial hyperplane H (with outward normal $\boldsymbol{\eta}$, say) supporting $\mathscr{R}(t^*)$ at $\mathbf{y}_1$. In other words

$$\boldsymbol{\eta}^{\mathrm{T}}\mathbf{y}_1 \geq \boldsymbol{\eta}^{\mathrm{T}}\mathbf{y} \qquad \text{for all} \quad \mathbf{y} \in \mathscr{R}(t^*) \quad \text{and} \quad \boldsymbol{\eta} \neq \mathbf{0},$$

that is

$$\boldsymbol{\eta}^{\mathrm{T}}\left(\int_0^{t^*} y_1(\tau)u^*(\tau)\,d\tau, \ldots, \int_0^{t^*} y_n(\tau)u^*(\tau)\,d\tau\right)$$

$$\geq \boldsymbol{\eta}^{\mathrm{T}}\left(\int_0^{t^*} y_1(\tau)u(\tau)\,d\tau, \ldots, \int_0^{t^*} y_n(\tau)u(\tau)\,d\tau\right) \qquad \text{for all} \quad u \in \Delta.$$

Rearranging gives (with $\boldsymbol{\eta}^{\mathrm{T}} = (\eta_1, \ldots, \eta_n)$)

$$\int_0^{t^*} (\eta_1 y_1(\tau) + \eta_2 y_2(\tau) + \cdots + \eta_n y_n(\tau))(u^*(\tau) - u(\tau))\,d\tau \geq 0, \qquad \text{for all } u \in \Delta,$$

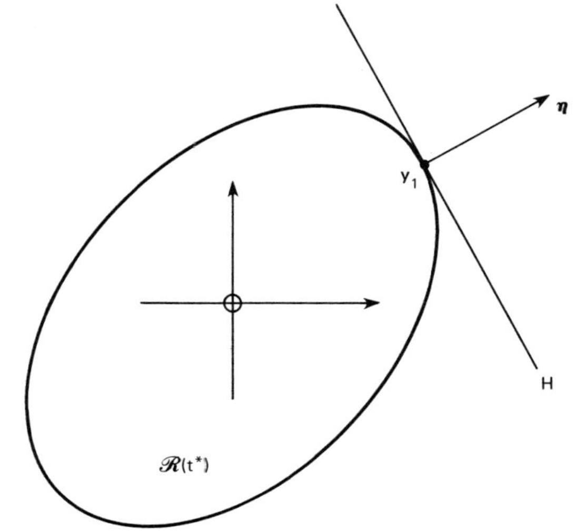

Fig. 2

or

$$\int_0^{t^*} (\boldsymbol{\eta}^{\mathrm{T}}\mathbf{y}(\tau))(u^*(\tau)-u(\tau))\,d\tau \geq 0.$$

This can happen only if $u^*(\tau) = \operatorname{sgn}(\boldsymbol{\eta}^{\mathrm{T}}\mathbf{y}(\tau))$, $\tau \in (0, t^*)$. Here

$$\operatorname{sgn}(x) = \begin{cases} +1 & \text{if} \quad x > 0 \\ -1 & \text{if} \quad x < 0 \\ \text{undefined} & \text{if} \quad x = 0. \end{cases}$$

Hence we have just proven the following theorem, a special case of the *Pontryagin maximum principle* [6].

Theorem 1 If u^* is any optimal control transfering (1) from $\mathbf{x}_0$ to $\mathbf{x}_1$ in minimum time t^*, then there exists a nonzero $\boldsymbol{\eta} \in R^n$ such that

$$u^*(t) = \operatorname{sgn}(\boldsymbol{\eta}^{\mathrm{T}}\mathbf{y}(t)) = \operatorname{sgn}(\boldsymbol{\eta}^{\mathrm{T}}X^{-1}(t)B(t)), \qquad t \in [0, t^*]. \tag{4}$$

In the general case, with $\mathbf{u}^* = (u_1^*, \ldots, u_m^*)^{\mathrm{T}}$, we can similarly show

$$u_i^*(t) = \operatorname{sgn}([\boldsymbol{\eta}^{\mathrm{T}}X^{-1}(t)B(t)]_i) \tag{5}$$

for $i = 1, \ldots, m$, $t \in [0, t^*]$, where $[\ ]_i$ denotes the ith component of the vector. To simplify the notation we shall henceforth abbreviate (5) by

$$\mathbf{u}^*(t) = \operatorname{sgn}(\boldsymbol{\eta}^{\mathrm{T}}X^{-1}(t)B(t))^{\mathrm{T}}, \qquad t \in [0, t^*]. \tag{6}$$

We can place this theorem in the form it is usually stated by noting that the function $\psi(t) = \eta^T X^{-1}(t)$ is the solution of the *adjoint equation*

$$\dot{\psi}(t) = -\psi(t)A(t)$$

with initial condition $\psi(0) = \eta^T$. For, taking inverses and differentiating gives

$$0 = \frac{d}{dt}(\eta^T) = \frac{d}{dt}(\psi(t)X(t)) = \dot{\psi}X + \psi\dot{X}$$
$$= \dot{\psi}X + \psi AX = (\dot{\psi} + \psi A)X$$

and consequently, $\dot{\psi} = -\psi A$. Further $\psi(0) = \eta^T$ since $X(0) = I$. Relations (4) and (6) then become

$$\mathbf{u}^*(t) = \operatorname{sgn}(\psi(t)B(t))^T, \qquad 0 \le t \le t^*. \tag{7}$$

If we define the *Hamiltonian* by

$$H(\psi, \mathbf{x}, \mathbf{u}) = \psi(A\mathbf{x} + B\mathbf{u}),$$

then the optimal control (7) satisfies

$$H(\psi, \mathbf{x}, \mathbf{u}^*) = \max_{\mathbf{u} \in \Delta} H(\psi, \mathbf{x}, \mathbf{u}).$$

We shall return later to an in-depth discussion of the maximum principle, but first we shall see how the characterization (7) allows us to determine time optimal controls. One immediate observation is that $\mathbf{u}^*$ will be bang-bang whenever $\eta^T X^{-1}(t)B(t)$ is nonzero, and the times at which the control changes value are just the zeros of this function. These times are usually called the *switching times* of $\mathbf{u}^*$.

2. APPLICATIONS OF THE MAXIMUM PRINCIPLE

Consider the control problem in Example 1 in Chapter I. The state equation is

$$\dot{\mathbf{x}}(t) = \begin{bmatrix} 0 & 1 \\ 0 & 0 \end{bmatrix} \mathbf{x}(t) + \begin{bmatrix} 0 \\ 1 \end{bmatrix} u(t), \qquad \mathbf{x}(0) = (x_0, y_0). \tag{1}$$

Hence

$$A = \begin{bmatrix} 0 & 1 \\ 0 & 0 \end{bmatrix}, \qquad B = \begin{bmatrix} 0 \\ 1 \end{bmatrix},$$

and $X(t)$ is the solution of

$$\dot{X}(t) = AX(t), \qquad X(0) = I.$$

Then

$$X(t) = e^{At} = I + \sum_{n=1}^{\infty} \frac{A^n t^n}{n!}$$

and so

$$X^{-1}(t) = e^{-At} = \begin{bmatrix} 1 & -t \\ 0 & 1 \end{bmatrix}.$$

Consequently, from the maximum principle the optimal control has the form

$$u^*(t) = \operatorname{sgn}(\eta_1 t + \eta_2) \qquad \text{for some} \quad (\eta_1, \eta_2) \in R^2. \tag{2}$$

We see immediately that u^* has only one switch between 1 and -1 (as we conjectured). Note that when $u^* = +1$ we can solve (1) with initial condition $\mathbf{x}(0) = (x_0, y_0)$ and obtain

$$\frac{dx_1}{dt} = x_2, \qquad \frac{dx_2}{dt} = 1, \qquad \text{i.e.,} \quad \frac{dx_2}{dx_1} = \frac{1}{x_2}, \quad \mathbf{x}(0) = (x_0, y_0),$$

$$x_1 = \frac{x_2^2}{2} + x_0 - \frac{y_0^2}{2}$$

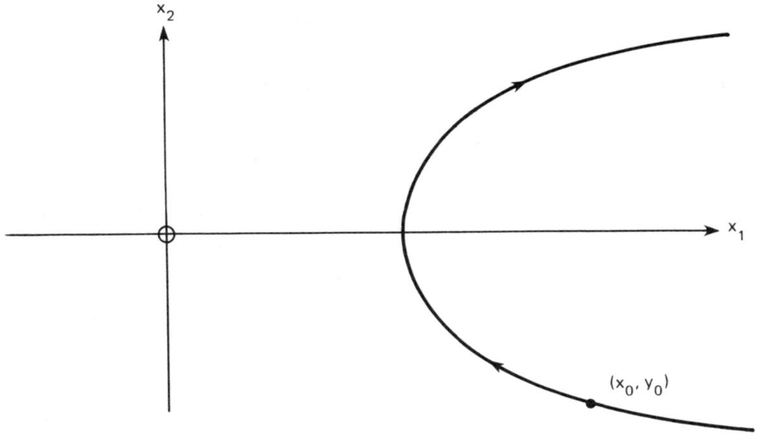

Fig. 3

(see Fig. 3). Note that x_2 is always increasing, so we traverse the path in the direction indicated.

When $u = -1$, we have

$$x_1 = -\tfrac{1}{2}x_2^2 + (x_0 + \tfrac{1}{2}y_0^2),$$

which is shown in Fig. 4. This time x_2 is decreasing.

The maximum principle tells us that the optimal control is either $u^* = +1$ (then $u^* = -1$), or $u^* = -1$ (then $u^* = +1$). So, bearing in mind that we have to end up at **0** by traversing arcs of the types in Figs. 3 and 4, we can easily construct the optimal trajectory (see Fig. 5).

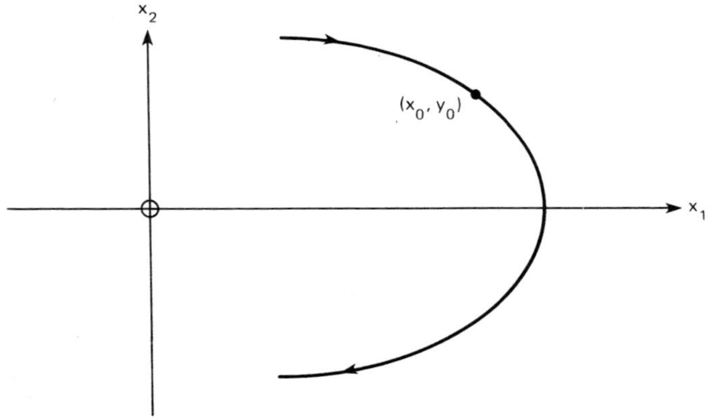

Fig. 4

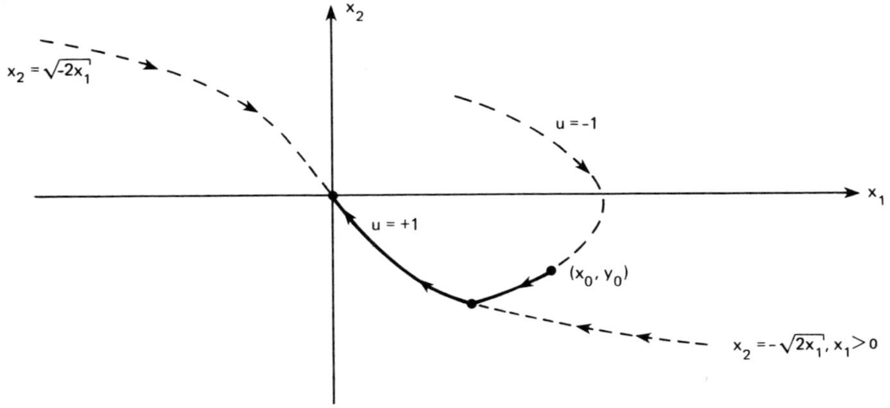

Fig. 5

For example, if $x_0 = 4$ and $y_0 = -1$, then we must control first with $u = -1$, and when the two parabolas in Fig. 5 intersect we change to $u = +1$. To calculate the first switch t_1, we find the equation of the first parabola:

$$x_1 = \tfrac{1}{2}(9 - x_2^2)$$

or $x_2 = -\sqrt{9 - 2x_1}$ (since $x_2 < 0$). Hence the point of intersection occurs at

$$-\sqrt{9 - 2x_1} = -\sqrt{2x_1},$$

i.e.,

$$9 - 2x_1 = 2x_1,$$

$$x_1 = \frac{9}{4}, \qquad x_2 = -\sqrt{9 - \frac{9}{2}} = -\frac{3}{\sqrt{2}} = -\frac{3\sqrt{2}}{2}.$$

Now solving the state equations (with $u = -1$) in terms of t gives

$$x_2(t) = -t + y_0 = -3\sqrt{2}/2,$$

so

$$t_1 + 1 = \frac{3\sqrt{2}}{2}, \qquad t_1 = \frac{3\sqrt{2} - 2}{2} = 1.121\ldots$$

is the switching time.

Similarly, after changing to $u = +1$, we traverse

$$\dot{x}_1 = x_2, \qquad \dot{x}_2 = 1$$

from $(\frac{9}{4}, -3\sqrt{2}/2)$ to $(0, 0)$. Solving for x_2, we find $x_2(t) = t - 3\sqrt{2}/2$, and $x_2 = 0$ when $t = 3\sqrt{2}/2$ sec., that is, the total time taken is $t^* = 2(3\sqrt{2}/2) - 1 = 3\sqrt{2} - 1 = 3.242$ sec.

We can formalize this argument by defining the following curve, called the *switching locus* Γ:

$$x_2 = W(x_1) = \begin{cases} -\sqrt{2x_1} & \text{for } x_1 \geq 0 \\ +\sqrt{-2x_1} & \text{for } x_1 < 0 \end{cases}$$

(see Fig. 6). Then if we set

$$\psi(x_1, x_2) = \begin{cases} -1 & \text{if } x_2 > W(x_1) \text{ or if } (x_1, x_2) \in \Gamma_- \\ 0 & \text{if } x_1 = x_2 = 0 \\ 1 & \text{if } x_2 < W(x_1) \text{ or if } (x_1, x_2) \in \Gamma_+, \end{cases}$$

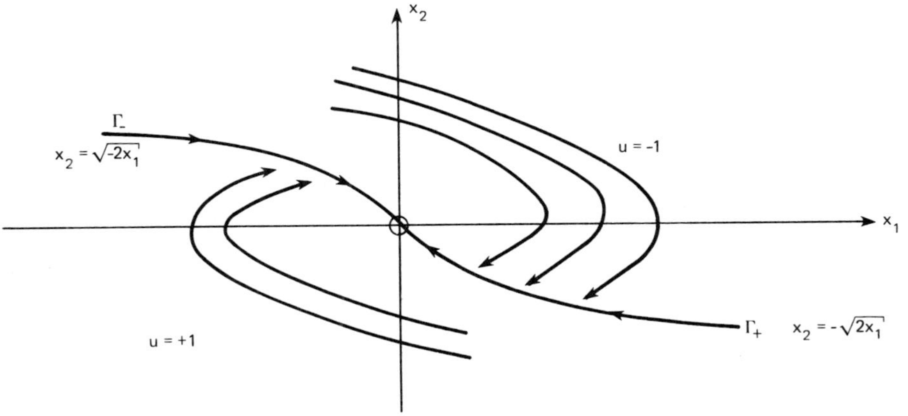

Fig. 6. Switching locus $\Gamma = \Gamma_- \cup \Gamma_+$.

the optimum trajectory starting from $\mathbf{x}(0) = (x_0, y_0)$ is just the solution of

$$\ddot{z} = \psi(z, \dot{z}), \qquad z(0) = x_0, \qquad \dot{z}(0) = y_0,$$

which terminates at $z(t^*) = 0$, $\dot{z}(t^*) = 0$. Given any point $(z(t), \dot{z}(t))$ on this trajectory, the value of the optimal control at this time t is

$$u^*(t) = \psi(z(t), \dot{z}(t)).$$

With the equation written in this way, an observer placed on the trolley will be able to operate the thrust optimally knowing only his position and velocity at every time t. Controls of this type are called *closed loop* or *feedback controls.*

3. NORMAL SYSTEMS—UNIQUENESS OF THE OPTIMAL CONTROL

For this section, we shall suppose *A and B are constant matrices.* Then $X(t) = e^{At}$ and the maximum principle states that optimal controls have the form

$$\mathbf{u}^*(t) = \operatorname{sgn}(\boldsymbol{\eta}^{\mathrm{T}} e^{-At} B)^{\mathrm{T}}, \qquad t \in [0, t^*], \tag{1}$$

for some $\boldsymbol{\eta} \neq \mathbf{0}$ in R^n. Writing $B = (\mathbf{b}_1, \ldots, \mathbf{b}_m)$ in columns, we can state this equivalently,

$$u_j^*(t) = \operatorname{sgn}(\boldsymbol{\eta}^{\mathrm{T}} e^{-At} \mathbf{b}_j), \qquad t \in [0, t^*], \tag{2}$$

$j = 1, 2, \ldots, m$. The only way the maximum principle will not determine the optimal control uniquely is if, for some $j = 1, 2, \ldots, m$, the function

$$t \mapsto \boldsymbol{\eta}^{\mathrm{T}} e^{-At} \mathbf{b}_j, \qquad t \in [0, t^*],$$

is zero on some interval. Of course systems for which the maximum principle has only one solution for each $\boldsymbol{\eta} \neq \mathbf{0}$, and so gives the optimal control uniquely, deserve a special name.

Definition 1 (Hermes and La Salle [3]) We call the control system (1.1) (with A and B constant) *normal* if for every nonzero $\boldsymbol{\eta} \in R^n$, the functions $t \mapsto \boldsymbol{\eta}^{\mathrm{T}} e^{-At} \mathbf{b}_j$, $t \in [0, t^*]$, $j = 1, 2, \ldots, m$, are nonzero, except at a finite number of points. Then, *if* (1) *is normal, the optimal control is unique, bang-bang with a finite number of switches, and is given by* (1) (*or* (2)) *for some* $\boldsymbol{\eta} \in R^n$.

At the end of this section we shall derive a simple necessary and sufficient test for normality, but first we look at the geometric meaning of normality.

Reconsidering our proof of the maximum principle, we showed that $\boldsymbol{\eta}$ was the normal to a hyperplane H, which supported $\mathscr{R}(t^*)$ at $\mathbf{y}_1$ (see Fig. 7).

If H happens to touch $\mathscr{R}(t^*)$ at another point $\mathbf{y}_2$, say, then by definition

$$\boldsymbol{\eta}^{\mathrm{T}} \mathbf{y}_2 \geq \boldsymbol{\eta}^{\mathrm{T}} \mathbf{y} \qquad \text{for all} \quad \mathbf{y} \in \mathscr{R}(t^*).$$

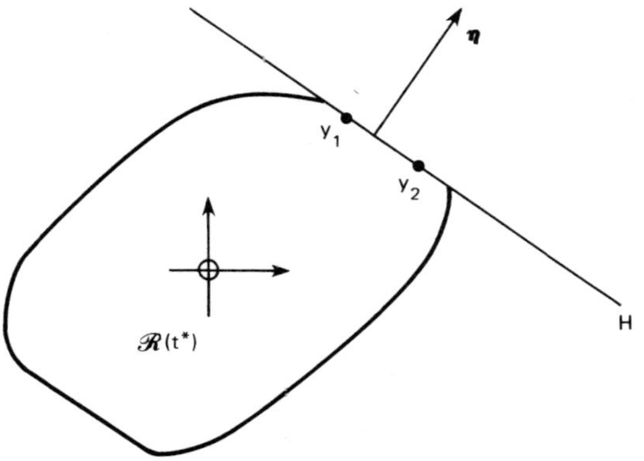

Fig. 7

If $\mathbf{y}_2$ is reached by control $\mathbf{u}_2$ in time t^*, that is,

$$\mathbf{y}_2 = \int_0^{t^*} e^{-At} B\mathbf{u}_2(t)\, dt,$$

then, as in the proof of the maximum principle,

$$\mathbf{u}_2(t) = \operatorname{sgn}(\boldsymbol{\eta}^{\mathrm{T}} e^{-At} B), \qquad t \in [0, t^*],$$

or $\mathbf{u}_2$ is another solution of the maximum principle ($\mathbf{u}_2 \not\equiv \mathbf{u}^*$, otherwise $\mathbf{y}_1 = \mathbf{y}_2$). In other words, any control that steers us in time t^* to a point in $H \cap \mathscr{R}(t^*)$ satisfies the maximum principle *for the same* $\boldsymbol{\eta}$. Equivalently, the maximum principle gives information only about the intersection of the hyperplane H (with normal $\boldsymbol{\eta}$) with the set $\mathscr{R}(t^*)$. The larger this set is (for instance if $\mathscr{R}(t^*)$ has a "flat"), the less information about our particular optimal control $\mathbf{u}^*$ we can derive from it.

For normal systems, the maximum principle tells us everything about $\mathbf{u}^*$, so the set $H \cap \mathscr{R}(t^*)$ must be as small as possible; in fact, $H \cap \mathscr{R}(t^*) = \{\mathbf{y}_1\}$. In our geometric language we can state this as a theorem:

Theorem 1 The optimal control $\mathbf{u}^*$ is uniquely determined by the maximum principle if and only if $\mathbf{y}_1$ is an exposed point of $\mathscr{R}(t^*)$.

We leave as an exercise the following:

Corollary The system (1.1) is normal if and only if $\mathscr{R}(t)$ is strictly convex for all $t > 0$.

As promised, we shall now derive a simple test for normality.

Theorem 2 The control system is normal if and only if for each $j = 1, 2, \ldots, m$, the vectors $\{\mathbf{b}_j, A\mathbf{b}_j, A^2\mathbf{b}_j, \ldots, A^{n-1}\mathbf{b}_j\}$ are linearly independent.

Proof Suppose $\boldsymbol{\eta}$ is an arbitrary nonzero vector in R^n. If the function

$$t \mapsto \boldsymbol{\eta}^{\mathrm{T}} e^{-At} \mathbf{b}_j, \qquad t \in [0, t^*], \tag{3}$$

is zero at more than a finite number of points, it must be identically zero, $\boldsymbol{\eta}^{\mathrm{T}} e^{-At} \mathbf{b}_j = 0$ for all $t \in [0, t^*]$, as (3) is an analytic function. Hence if (1.1) is not normal for some j, we must have $\boldsymbol{\eta}^{\mathrm{T}} e^{-At} \mathbf{b}_j \equiv 0$.

Substituting $t = 0$ gives $\boldsymbol{\eta}^{\mathrm{T}} \mathbf{b}_j = 0$. Differentiating (3) gives

$$\frac{d}{dt}(\boldsymbol{\eta}^{\mathrm{T}} e^{-At} \mathbf{b}_j) = -\boldsymbol{\eta}^{\mathrm{T}} A e^{-At} \mathbf{b}_j \equiv 0$$

and, at $t = 0$,

$$\boldsymbol{\eta}^{\mathrm{T}} A \mathbf{b}_j = 0.$$

Similarly we can show

$$\boldsymbol{\eta}^{\mathrm{T}} A^r \mathbf{b}_j = 0 \qquad \text{for all} \quad r = 0, 1, 2, \ldots, n-1,$$

and so $\{\mathbf{b}_j, A\mathbf{b}_j, \ldots, A^{n-1}\mathbf{b}_j\}$ must be linearly dependent, that is, *linear independence implies normality.*

To prove the converse, assume that for some j, the vectors $\{\mathbf{b}_j, A\mathbf{b}_j, \ldots, A^{n-1}\mathbf{b}_j\}$ are linearly dependent. Then there exists a nonzero $\boldsymbol{\eta} \in R^n$ with

$$\boldsymbol{\eta}^{\mathrm{T}}\mathbf{b}_j = \boldsymbol{\eta}^{\mathrm{T}} A\mathbf{b}_j = \cdots = \boldsymbol{\eta}^{\mathrm{T}} A^{n-1}\mathbf{b}_j = 0. \tag{4}$$

Define

$$\xi(t) = \boldsymbol{\eta}^{\mathrm{T}} e^{-At}\mathbf{b}_j;$$

then

$$(D^k\xi)(t) = \boldsymbol{\eta}^{\mathrm{T}}(-A)^k e^{-At}\mathbf{b}_j,$$

where $D \equiv d/dt$. We shall now show that $\xi \equiv 0$. Let $\varphi(-\lambda)$ be the characteristic polynomial of A. By the properties of matrix exponentials,

$$\varphi(D)\xi = \varphi(D)[\boldsymbol{\eta}^{\mathrm{T}} e^{-At}\mathbf{b}_j] = \boldsymbol{\eta}^{\mathrm{T}}[\varphi(D)e^{-At}\mathbf{b}_j] = \boldsymbol{\eta}^{\mathrm{T}}[\varphi(-A)e^{-At}\mathbf{b}_j].$$

However, by the *Hamilton–Cayley theorem* [7], $\varphi(-A) = 0$, and so

$$(\varphi(D)\xi)(t) = 0 \qquad \text{for all} \quad t \in [0, t^*]. \tag{5}$$

Further by (4),

$$[D^k\xi](0) = 0, \qquad k = 0, 1, 2, 3, \ldots, n-1. \tag{6}$$

Equations (5) and (6) are just a linear homogeneous ordinary differential equation for the function ξ, with zero initial data, and, by the uniqueness of the solutions for such problems, we must have $\xi(t) = 0$ for all $t \in [0, t^*]$, which contradicts the definition of normality. ∎

4. FURTHER EXAMPLES OF TIME OPTIMAL CONTROL

Example 1 Consider the control problem

$$\begin{aligned} \dot{x}_1 &= x_2, \\ \dot{x}_2 &= -x_1 + u, \qquad x(0) = (x_0, y_0), \qquad |u| \le 1. \end{aligned} \tag{1}$$

We wish to reach the origin in minimum time.

We have

$$\begin{bmatrix} \dot{x}_1 \\ \dot{x}_2 \end{bmatrix} = \begin{bmatrix} 0 & 1 \\ -1 & 0 \end{bmatrix} \begin{bmatrix} x_1 \\ x_2 \end{bmatrix} + \begin{bmatrix} 0 \\ 1 \end{bmatrix} u$$

i.e.,

$$A = \begin{bmatrix} 0 & 1 \\ -1 & 0 \end{bmatrix}, \qquad \mathbf{b} = \begin{bmatrix} 0 \\ 1 \end{bmatrix}, \qquad A\mathbf{b} = \begin{bmatrix} 1 \\ 0 \end{bmatrix};$$

hence $\{\mathbf{b}, A\mathbf{b}\}$ are linearly independent and the system is normal. Next

$$X(t) = e^{At} = \begin{bmatrix} \cos t & \sin t \\ -\sin t & \cos t \end{bmatrix}, \qquad e^{-At}\mathbf{b} = \begin{bmatrix} -\sin t \\ \cos t \end{bmatrix}.$$

By the maximum principle, optimal controls must be of the form

$$u^*(t) = \operatorname{sgn}(-\eta_1 \sin t + \eta_2 \cos t) \qquad (\eta_1, \eta_2) \neq (0, 0)$$

or

$$u^*(t) = \operatorname{sgn}(\sin(t + \delta)) \qquad (\tan \delta = -\eta_2/\eta_1)$$

for some $-\pi \leq \delta \leq \pi$. That is, the optimal control is *unique, bang-bang,* and its *switches* occur *exactly* π *seconds apart.* When $u = +1$,

$$\begin{aligned} \dot{x}_1 &= x_2, \\ \dot{x}_2 &= -x_1 + 1, \\ \frac{dx_1}{dx_2} &= \frac{x_2}{-x_1 + 1} \\ 1 - x_1 &= a \cos t, \\ x_2 &= a \sin t, \end{aligned}$$

and

$$\begin{aligned} (1 - x_1)\, dx_1 &= x_2 dx_2, \\ x_2^2 + (1 - x_1)^2 &= a^2, \end{aligned}$$

which is a circle centered at $(1, 0)$. Similarly, when $u = -1$,

$$x_2^2 + (-1 - x_1)^2 = a^2,$$

which is a circle centered at $(-1, 0)$, where

$$\begin{aligned} x_1 &= a \cos t - 1, \\ x_2 &= -a \sin t. \end{aligned}$$

With t increasing, these circles are transversed in a clockwise sense; with t decreasing, in an counter clockwise sense. To solve this control problem, suppose we start at the origin and move backward in time until we hit (x_0, y_0) in time $-t^*$. Since we are moving backward in time, we transverse the circles counterclockwise.

Suppose first that $0 < \delta \leq \pi$ (see Fig. 8). Then we move $-\delta$ seconds around the arc of the trajectory corresponding to $u = +1$, which passes through the origin. *At* $t = -\delta$, $\sin(t + \delta)$ changes sign and we switch the control to $u = -1$. Since $u = -1$, the optimal trajectory is circular with center at $(-1, 0)$, and passing through P_1. We travel along this circle for π seconds, in which time we traverse exactly a semicircle. (From Fig. 8 we see that each circle is traversed in 2π seconds.) After π seconds we shall reach the point P_2, which by symmetry is just the reflection of P_1 onto the circle with radius 1 and center $(-1, 0)$. At P_2 we switch to $u = +1$ again and traverse a semicircle which has a center at $(1, 0)$ and which passes through P_2. After π seconds we reach P_3 and switch to $u = -1$, etc.

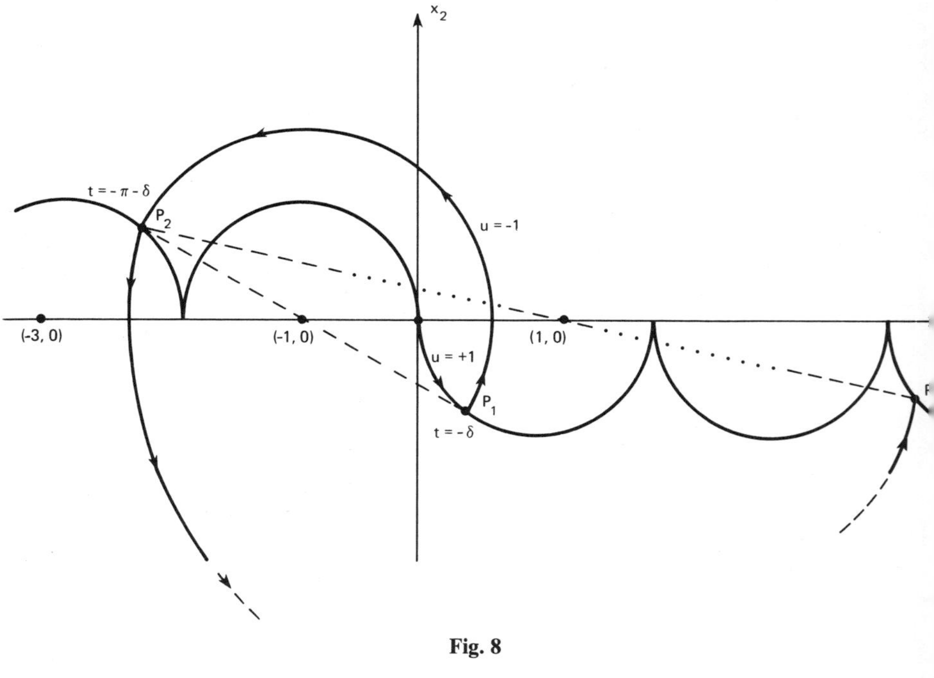

Fig. 8

In this way we generate the optimum trajectories, and the one that passes through (x_0, y_0) must be (by normality) the desired optimal trajectory.

If $-\pi \le \delta < 0$, then we switch with $u = -1$ until $\sin(t+\delta) = 0$ i.e., $t = -\delta - \pi$ seconds (see Fig. 9). We then switch to $u = +1$, describing a semicircular arc with center $(1, 0)$ for π seconds to Q_2, switch to $u = -1$, etc. Clearly, the switching locus W is just as shown in Fig. 10, and the synthesizer is defined by

$$\psi(x, y) = \begin{cases} -1 & \text{if } (x, y) \text{ lies above } W \text{ or on } \Gamma_- \\ +1 & \text{if } (x, y) \text{ lies below } W \text{ or on } \Gamma_+. \end{cases}$$

The optimal responses are just the solutions of

$$\ddot{x}_1 + x_1 = \psi(x_1, x_2) = \psi(x_1, \dot{x}_1)$$

with initial point (x_0, y_0) and final point $(0, 0)$.

Example 2 (Bushaw [1]; Lee and Marcus [5]) Consider the minimal time control to the origin for

$$\ddot{x} + 2b\dot{x} + k^2 x = u, \qquad x(0) = x_0, \qquad \dot{x}(0) = y_0 \tag{2}$$

(the *damped linear oscillator*) where $b > 0$ and $k > 0$ are constants and $|u(t)| \le 1$. First, (2) is equivalent to

$$\begin{bmatrix} \dot{x}_1 \\ \dot{x}_2 \end{bmatrix} = \begin{bmatrix} 0 & 1 \\ -k^2 & -2b \end{bmatrix} \begin{bmatrix} x_1 \\ x_2 \end{bmatrix} + \begin{bmatrix} 0 \\ 1 \end{bmatrix} u, \qquad \mathbf{x}(0) = (x_0, y_0). \tag{3}$$

To begin with, we shall suppose $b^2 - k^2 \ge 0$; (2) is then *critically* or *over damped.*

The maximum principle becomes (from the alternative formulation in terms of the adjoint system)

$$\dot{\boldsymbol{\psi}} = -\boldsymbol{\psi} A,$$
$$u^*(t) = \operatorname{sgn}(\boldsymbol{\psi}(t)\mathbf{b}) = \operatorname{sgn}(\psi_2(t)).$$

Writing this out, we see that

$$(\dot{\psi}_1, \dot{\psi}_2) = (\psi_1, \psi_2) \begin{bmatrix} 0 & -1 \\ k^2 & 2b \end{bmatrix}$$

that is, $\ddot{\psi}_2 - 2b\dot{\psi}_2 + k^2\psi_2 = 0$, which has solutions

$$\psi_2(t) = \begin{cases} e^{bt}(\alpha + \beta t) & \text{if } b^2 - k^2 = 0 \\ \alpha e^{bt} \sinh(\mu t + \beta) & \text{if } b^2 - k^2 > 0, \end{cases}$$

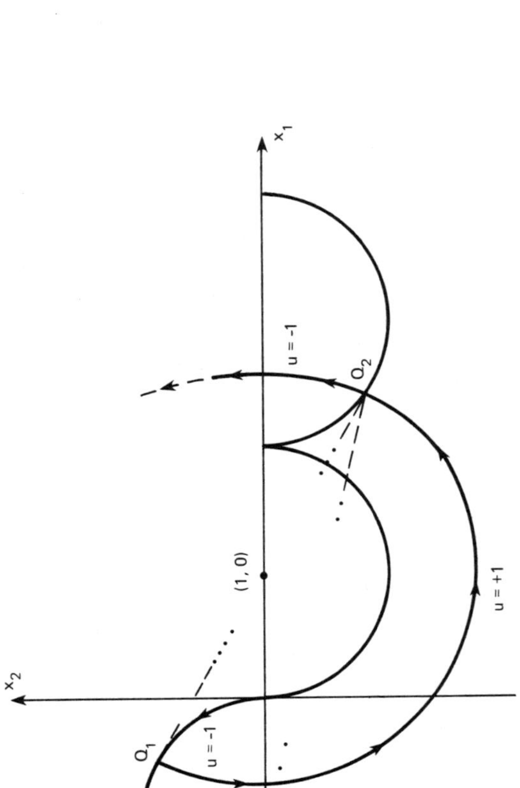

Fig. 9

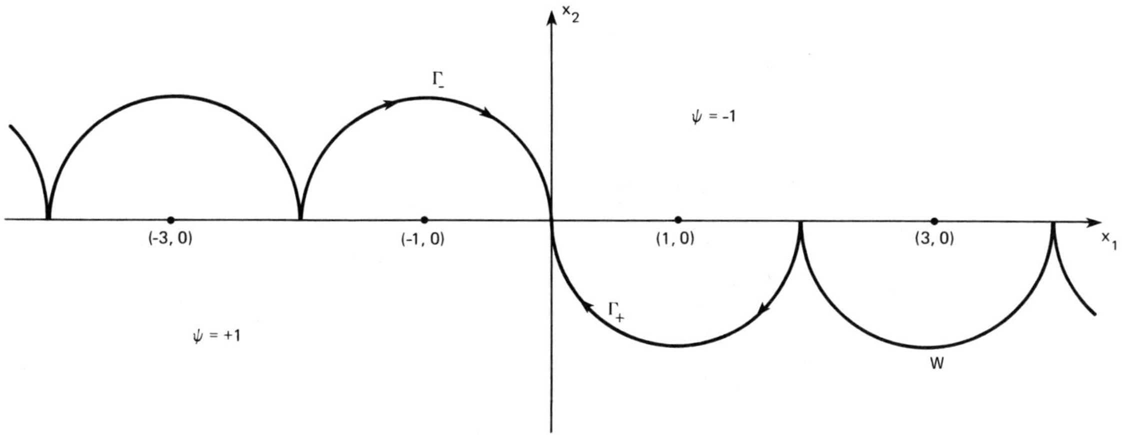

Fig. 10

where α, β are constants (initial conditions for the adjoint equation) and $\mu = \sqrt{b^2 - k^2}$. In any case, ψ_2 can have at most one zero, and u^* has at most one switch.

If we denote by

$$(\Gamma_+) \qquad \begin{bmatrix} \dot{x}_1 \\ \dot{x}_2 \end{bmatrix} = \begin{bmatrix} 0 & 1 \\ -k^2 & -2b \end{bmatrix} \begin{bmatrix} x_1 \\ x_2 \end{bmatrix} + \begin{bmatrix} 0 \\ 1 \end{bmatrix}$$

the solution of (3) passing through (0, 0) with $u = +1$ in the lower right-hand quadrant and by

$$(\Gamma_-) \qquad \begin{bmatrix} \dot{x}_1 \\ \dot{x}_2 \end{bmatrix} = \begin{bmatrix} 0 & 1 \\ -k^2 & -2b \end{bmatrix} \begin{bmatrix} x_1 \\ x_2 \end{bmatrix} - \begin{bmatrix} 0 \\ 1 \end{bmatrix}$$

the solution of (3) passing through (0, 0) with $u = -1$ in the upper left-hand quadrant, then the switching locus $x_2 = W(x_1)$ is as pictured in Fig. 11. The optimal control synthesizer is then

$$\psi(x_1, x_2) = \begin{cases} -1 & \text{for} \quad x_2 > W(x_1) \text{ and on } (\Gamma_-) \\ +1 & \text{for} \quad x_2 < W(x_1) \text{ and on } (\Gamma_+). \end{cases}$$

The verification of these details is exactly the same as in Section 2 Example 1.

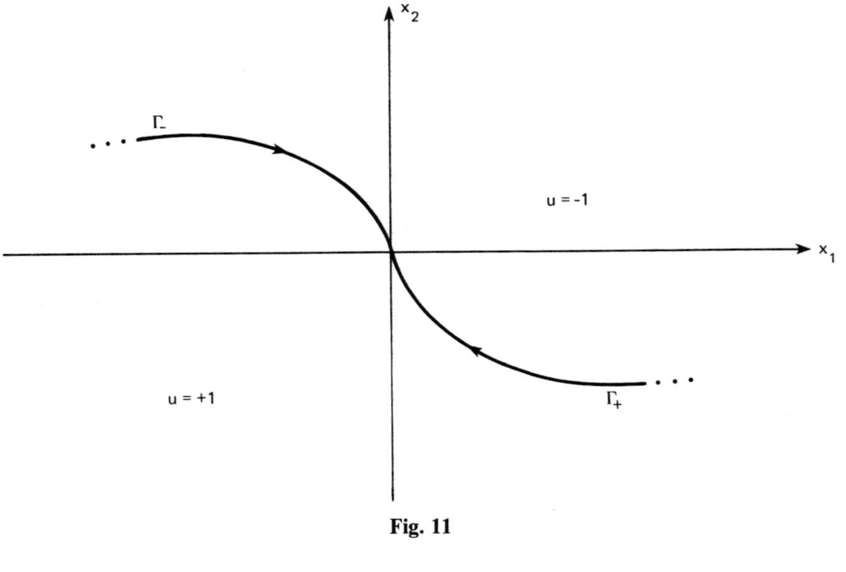

Fig. 11

Consider now the case of underdamping $b^2 - k^2 < 0$. We can solve the adjoint system, and we get

$$\psi_2(t) = \alpha e^{bt} \sin(\omega t + \beta),$$

where $\omega = \sqrt{k^2 - b^2}$. In other words, the switches of the optimal control are exactly π/ω seconds apart.

Each solution of the state equations (with $u = +1$)

$$(S_+) \qquad \begin{bmatrix} \dot{x}_1 \\ \dot{x}_2 \end{bmatrix} = \begin{bmatrix} 0 & 1 \\ -k^2 & -2b \end{bmatrix} \begin{bmatrix} x_1 \\ x_2 \end{bmatrix} + \begin{bmatrix} 0 \\ 1 \end{bmatrix}$$

is a spiral approaching a critical equilibrium point $O_+ = (1/k^2, 0)$ as $t \to +\infty$ (see Fig. 12). Similarly for $u = -1$, each solution of

$$(S_-) \qquad \begin{bmatrix} \dot{x}_1 \\ \dot{x}_2 \end{bmatrix} = \begin{bmatrix} 0 & 1 \\ -k^2 & -2b \end{bmatrix} \begin{bmatrix} x_1 \\ x_2 \end{bmatrix} - \begin{bmatrix} 0 \\ 1 \end{bmatrix}$$

is a spiral approaching $O_- = (-1/k^2, 0)$ as $t \to \infty$ (see Fig. 13).

We construct the switching locus as follows: Find the solution of S_+ passing through the origin and unfold it, that is, take S_1 and reflect it back to $\hat{S}_1$, S_2 to $\hat{S}_2$, etc. (see Fig. 14).

This defines the switching locus $x_2 = W(x_1)$ for $x_1 \geq 0$. For $x_1 < 0$, we set $W(x_1) = -W(-x_1)$. This gives us the result shown in Fig. 15. Then it can be shown [5] that if P_1 and P_2 are colinear with O_+, it takes exactly π/ω seconds to traverse the optimal trajectory (S_+) between P_1 and P_2, so as in the case of the undamped oscillator, we start off at our initial point (x_0, y_0), with $u = +1$ if it lies below W, $u = -1$ if above W, and continue until we hit W. Then the control switches sign, and by

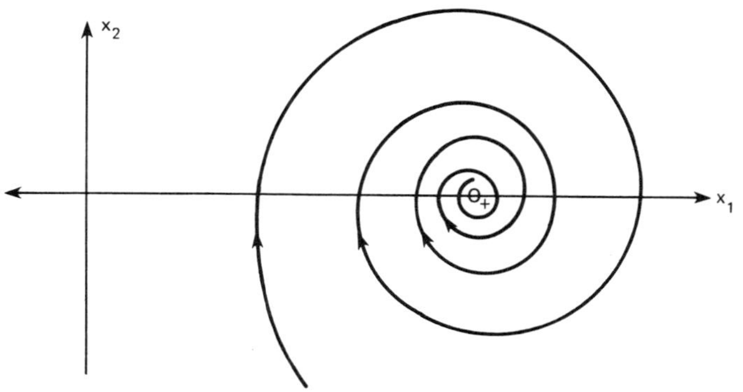

Fig. 12

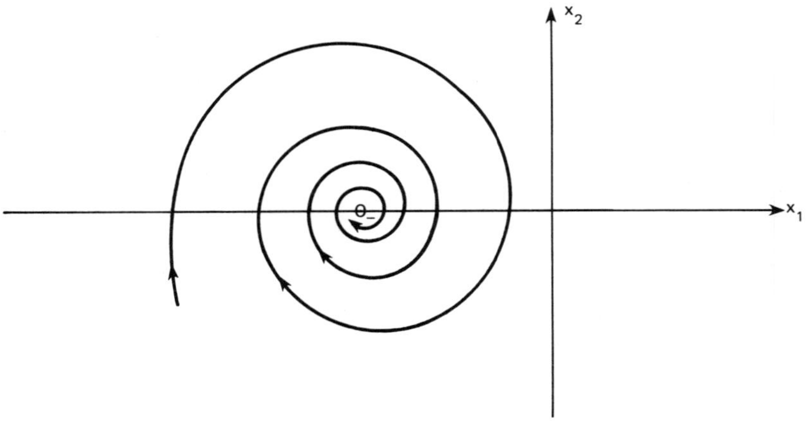

Fig. 13

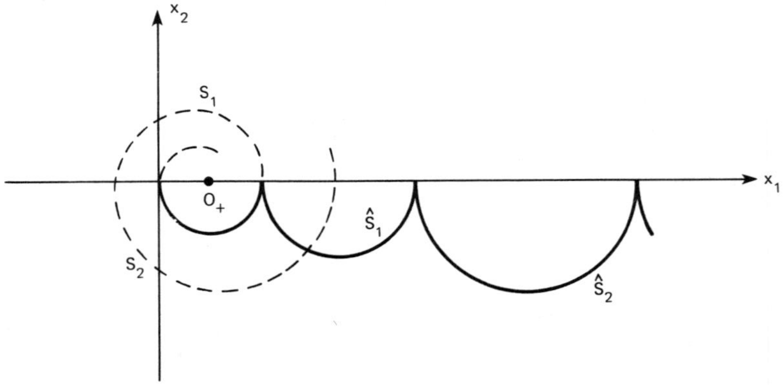

Fig. 14

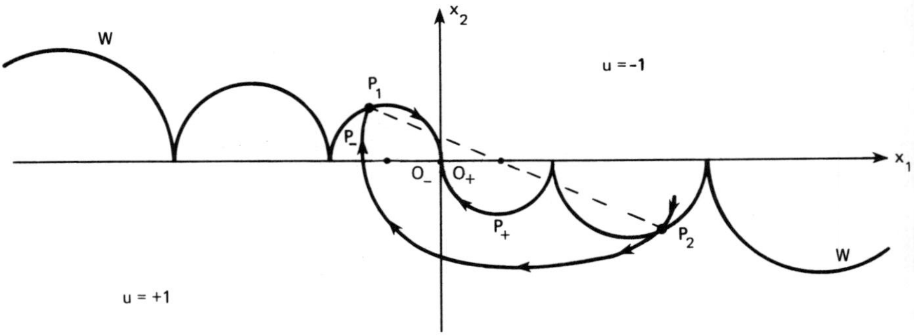

Fig. 15

the maximum principle we go for π/ω seconds with the control at this value. However, it takes us exactly π/ω seconds to rehit W, and so every time we cross W the control changes sign. This process continues until we hit Γ_- or Γ_+, and then we come into $\mathbf{0}$ by switching to $u = -1$ or $u = +1$, respectively.

The synthesizer can be defined by

$$\Psi(x_1, x_2) = \begin{cases} -1 & \text{for} \quad x_2 > W(x_1) \text{ and on } \Gamma_- \\ +1 & \text{for} \quad x_2 < W(x_1) \text{ and on } \Gamma_+ \end{cases}$$

and the optimal trajectories are solutions of

$$\ddot{x} + 2b\dot{x} + k^2 x = \psi(x, \dot{x}), \qquad x(0) = x_0, \qquad \dot{x}(0) = y_0.$$

5. NUMERICAL COMPUTATION OF THE SWITCHING TIMES

In this section we discuss a method for the numerical solution of time optimal control problems

$$\dot{\mathbf{x}} = A\mathbf{x} + \mathbf{b}u. \tag{1}$$

Suppose the optimal control u^* transferring $\mathbf{x}_0$ to $\mathbf{0}$ in minimum time is bang-bang with r switches. Without loss of generality, suppose that the first action of u^* is -1 (see Fig. 16). Then, we must have

$$\mathbf{0} = e^{+At_r}\left(\mathbf{x}_0 + \int_0^{t_r} e^{-At}\mathbf{b}u^*(t)\, dt\right)$$

i.e.,

$$-\mathbf{x}_0 = \int_0^{t_r} e^{-At}\mathbf{b}u^*(t)\, dt.$$

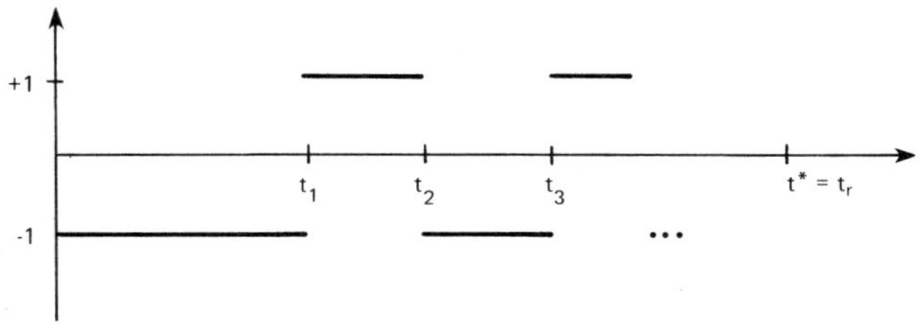

Fig. 16

In other words, multiplying both sides by A and integrating yields

$$A\mathbf{x}_0 = -(e^{-At_1}\mathbf{b} - \mathbf{b}) + (e^{-At_2}\mathbf{b} - e^{-At_1}\mathbf{b}) + \cdots + (-1)^r(e^{-At_r}\mathbf{b} - e^{-At_{r-1}}\mathbf{b})$$

i.e.,

$$A\mathbf{x}_0 = \mathbf{b} - 2e^{-At_1}\mathbf{b} + 2e^{-At_2}\mathbf{b} + \cdots + (-1)^r e^{-At_r}\mathbf{b}. \tag{2}$$

If matrix A is normal (commutes with its adjoint), has eigenvalues $\lambda_1, \ldots, \lambda_n$, and corresponding orthonormal eigenvectors $\{\mathbf{x}_1, \ldots, \mathbf{x}_n\}$ [7], then

$$e^{At}\mathbf{b} = \sum_{j=1}^{n} e^{\lambda_j t}(\mathbf{b} \cdot \mathbf{x}_j)\mathbf{x}_j$$

i.e.,

$$e^{-At}\mathbf{b} = \sum_{j=1}^{n} e^{-\lambda_j t}(\mathbf{b} \cdot \mathbf{x}_j)\mathbf{x}_j.$$

So we can expand (2) as

$$\sum \lambda_j(\mathbf{x}_0 \cdot \mathbf{x}_j)\mathbf{x}_j = \sum_{j=1}^{n} [(\mathbf{b} \cdot \mathbf{x}_j)\mathbf{x}_j - 2e^{-\lambda_j t_1}(\mathbf{b} \cdot \mathbf{x}_j)\mathbf{x}_j + 2e^{-\lambda_j t_2}(\mathbf{b} \cdot \mathbf{x}_j)\mathbf{x}_j + \cdots + (-1)^r e^{-\lambda_j t_r}(\mathbf{b} \cdot \mathbf{x}_j)\mathbf{x}_j]. \tag{3}$$

Now using the orthonormality of the $\{\mathbf{x}_j\}$, (3) can be alternatively written as a set of n nonlinear equations in r unknowns $t_1, \ldots, t_r$:

$$+\lambda_j(\mathbf{x}_0 \cdot \mathbf{x}_j) = (\mathbf{b} \cdot \mathbf{x}_j)(1 - 2e^{-\lambda_j t_1} + 2e^{-\lambda_j t_2} - 2e^{\lambda_j t_3} + \cdots + (-1)^r e^{-\lambda_j t_r})$$

for $j = 1, 2, \ldots, n$; that is,

$$-\lambda_1(\mathbf{x}_0 \cdot \mathbf{x}_1) + (\mathbf{b} \cdot \mathbf{x}_1)(1 - 2e^{-\lambda_1 t_1} + 2e^{-\lambda_1 t_2} + \cdots + (-1)^r e^{-\lambda_1 t_r}) = 0$$

$$\vdots$$

$$-\lambda_n(\mathbf{x}_0 \cdot \mathbf{x}_n) + (\mathbf{b} \cdot \mathbf{x}_n)(1 - 2e^{-\lambda_n t_1} + 2e^{-\lambda_n t_2} + \cdots + (-1)^r e^{-\lambda_n t_r}) = 0.$$

Note. It has been shown by Feldbaum [6, Chap. 3, Theorem 10] that if $\{\lambda_1, \ldots, \lambda_n\}$ are real, then $r \leq n - 1$.

Take as an example the harmonic oscillator, Example 1, Section 4,

$$\dot{\mathbf{x}} = A\mathbf{x} + \mathbf{b}u, \qquad \text{where} \quad \mathbf{x}(0) = \begin{bmatrix} x_0 \\ y_0 \end{bmatrix}.$$

Here we have

$$A = \begin{bmatrix} 0 & 1 \\ -1 & 0 \end{bmatrix} \quad \text{and} \quad \mathbf{b} = \begin{bmatrix} 0 \\ 1 \end{bmatrix}.$$

Hence

$$A\mathbf{x}_0 = \begin{bmatrix} y_0 \\ -x_0 \end{bmatrix}$$

and from Example 1,

$$e^{-At}\mathbf{b} = \begin{bmatrix} -\sin t \\ \cos t \end{bmatrix}.$$

Then (2) becomes

$$\begin{bmatrix} y_0 \\ -x_0 \end{bmatrix} = \begin{bmatrix} 0 \\ 1 \end{bmatrix} - 2\begin{bmatrix} -\sin t_1 \\ \cos t_1 \end{bmatrix} + 2\begin{bmatrix} -\sin t_2 \\ \cos t_2 \end{bmatrix} + \cdots + (-1)^r \begin{bmatrix} -\sin t_r \\ \cos t_r \end{bmatrix}$$

or

$$\sin t_1 - \sin t_2 + \cdots + \frac{(-1)^{r+1}}{2} \sin t_r = \frac{y_0}{2}$$

$$1 - 2\cos t_1 + 2\cos t_2 + \cdots + (-1)^r \cos t_r = -x_0.$$

These equations in general have multiple solutions; however, by normality, the problem

$$\text{minimize } t_r \qquad \text{subject to} \quad 0 < t_1 < t_2 < \cdots < t_r$$

and

$$f_1(t_1, \ldots, t_r) = \sin t_1 - \sin t_2 + \cdots + \frac{(-1)^{r+1}}{2} \sin t_r - \frac{y_0}{2} = 0,$$

$$f_2(t_1, \ldots, t_r) = 1 - 2\cos t_1 + 2\cos t_2 + \cdots + \frac{(-1)^r}{2} \cos t_r - x_0 = 0$$

must have a unique solution, namely, the *switching times of our optimal control* u^*. So, setting $\mathbf{t} = (t_1, \ldots, t_r)$, the problem has the general form

$$\text{P:} \quad \text{minimize } h(\mathbf{t}) \qquad \text{subject to} \quad g_i(\mathbf{t}) > 0 \quad \text{and} \quad f_j(\mathbf{t}) = 0, \quad j = 1, 2,$$
$$i = 1, \ldots, r.$$

This is a mathematical programming problem that could be solved by Lagrange multiplier methods. A better method is to make the following transformation:

$$t_1 = y_1^2,$$
$$t_2 = y_1^2 + y_2^2,$$
$$\vdots$$
$$t_r = y_1^2 + \cdots + y_r^2;$$

then setting $f_j^*(\mathbf{y}) = f_j(\mathbf{t})$, P is equivalent to

$$\mathrm{P^T}: \quad \text{minimize } y_1^2 + \cdots + y_n^2 \qquad \text{subject to} \quad f_j^*(\mathbf{y}) = 0, \quad j = 1, 2,$$

and we have removed the inequality constraints.

One method, which seems to work, for solving $\mathrm{P^T}$ is to solve

$$\mathrm{P^{TT}}: \quad \text{minimize } y_1^2 + \cdots + y_r^2 + A(f_1^*(\mathbf{y})^2 + f_2^*(\mathbf{y})^2),$$

increasing A until successive solutions $\mathbf{y}$ coincide to a desired accuracy (this is an example of a *penalty method*) [2, 3]. For example, with $x_0 = 1$, $y_0 = 0$, the switching times are $t_1 = 0.505$ sec, $t_2 = 1.823$ sec.

Problems

1. Consider a control process described by $\dot{x} + bx = u$ for a real constant b, with the restraint $|u(t)| \leq 1$. Verify that the response $x(t)$ with $x(0) = x_0$ to a control $u(t)$ is

$$x(t) = e^{-bt}x_0 + e^{-bt}\int_0^t e^{bs}u(s)\,ds.$$

(a) If $b \geq 0$, show that every initial point can be controlled to $x_1 = 0$.

(b) If $b < 0$, describe precisely those points x_0 that can be steered to $x_1 = 0$.

2. In the control process in Problem 1, show that the control transferring x_0 to $x_1 = 0$ in minimum time (when this is possible) is bang-bang, indicate how many switches it can have, and show that it can be synthesized by

$$u(t) = -\operatorname{sgn}(x(t)).$$

Compute the minimum time t in terms of x_0 and b.

3. Suppose that you have been given a contract to ship eggs by rocket from New York to Los Angeles, a distance of 2400 miles. Find the shortest time in which you can do this without breaking any eggs. You may assume the path traveled is a straight line, and neglect friction, the rotation of the earth, etc. The only stipulation is that the eggs break if the acceleration exceeds 100 ft/sec^2.
4. Calculate the minimum time to transfer from the initial point (1, 0) to the origin (0, 0) for the system

$$\ddot{x} + x = u, \qquad |u(t)| \leq 1.$$

What are the switching times of the optimal control?
5. Find the optimal trajectories and switching locus for the problem of reaching the origin in minimum time for the control system

$$\begin{aligned} \dot{x}_1 &= x_2 + u_1, & |u_1| &\leq 1, \\ \dot{x}_2 &= -x_1 + u_2, & |u_2| &\leq 1. \end{aligned}$$

6. Find the control that steers the system

$$\begin{aligned} \dot{x}_1 &= x_2, & x_1(0) &= 1, \\ \dot{x}_2 &= u, & x_2(0) &= 0, \end{aligned}$$

to the origin $(0, 0)$ in $t = 1$ and minimizes the cost

$$C(u) = \sup_{0 \leq t \leq 1} |u(t)|.$$

[*Hint*: If for the optimal control u^*, $C(u^*) = k$, show that this problem is equivalent to finding that number $k > 0$ for which the minimum time t^* to reach $(0, 0)$ from $(1/k, 0)$ for the system

$$\dot{x}_1 = x_2, \qquad \dot{x}_2 = u, \qquad |u| \leq 1$$

is $t^* = 1$.]
7. For the control system

$$\dot{x} = u - u^2, \qquad |u| \leq 1,$$

show that the minimum time control steering $x_0 = -1$ to $x_0 = 0$ is $u \equiv \frac{1}{2}$, which is not bang-bang.
8. Prove the Corollary to Theorem 3.1.
9. Discuss the time optimal control to the origin for the system,

$$\begin{aligned} \dot{x}_1 &= u_1 + u_2, & |u_1| &\leq 1, \\ \dot{x}_2 &= u_1 - u_2, & |u_2| &\leq 1. \end{aligned}$$

10. Show that the maximum of the Hamiltonian defined in Section 1 is constant in time if the control system $\dot{\mathbf{x}} = A\mathbf{x} + B\mathbf{u}$ is autonomous.

11. Prove that a point $\mathbf{y}_1 \in \mathscr{R}(t)$ is hit by a unique trajectory if and only if $\mathbf{y}_1$ is an extreme point of $\mathscr{R}(t)$.

References

[1] D. Bushaw, Optimal discontinuous forcing terms, *in* "Contributions to the Theory of Non-linear Oscillations," pp. 29–52. Princeton Univ. Press, Princeton, New Jersey, 1958.

[2] A. V. Fiacco and G. P. McCormick, "Non-linear Programming: Sequentially Unconstrained Minimization Techniques." Wiley, New York, 1968.

[3] H. Hermes and J. P. LaSalle, "Functional Analysis and Time-Optimal Control." Academic Press, New York, 1969.

[4] M. R. Hestenes, Multiplier and gradient methods, *J. Optim. Theory Appl.* **4**, 303–320 (1969).

[5] E. B. Lee and L. Markus, "Foundations of Optimal Control Theory." Wiley, New York, 1967.

[6] L. Pontryagin, V. Boltyanskii, R. Gramkrelidze, and E. Mischenko, "The Mathematical Theory of Optimal Processes." Wiley (Interscience), New York, 1962.

[7] G. Strang, "Computational Linear Algebra." Prentice-Hall, Englewood Cliffs, New Jersey, 1980.

Chapter III The Pontryagin Maximum Principle

1. THE MAXIMUM PRINCIPLE

Consider the autonomous control problem:

(1) $\dot{x}_i = f_i(x_1, \ldots, x_n, u_1, \ldots, u_m)$, $i = 1, 2, \ldots n$, with $\mathbf{f}$ continuously differentiable in $R^n \times \Omega$.

We are given

(2) the initial point $\mathbf{x}_0$ and (possibly) final point $\mathbf{x}_1$,

(3) the class Δ of admissible controls, which we take to be all piecewise continuous functions $\mathbf{u}$, with $\mathbf{u}(t) \in \Omega$, Ω a given set in R^m,

(4) the cost functional

$$C(\mathbf{u}) = \int_0^{t_1} f_0(\mathbf{x}(t), \mathbf{u}(t))\, dt,$$

where f_0 is continuously differentiable in $R^n \times \Omega$.

Define for $\mathbf{x}, \mathbf{u}, \boldsymbol{\varphi} \in R^n \times R^m \times R^n$, $\varphi_0 \in R$,

$$\begin{aligned} H(\mathbf{x}, \mathbf{u}, \boldsymbol{\varphi}) &= \varphi_0 f_0(\mathbf{x}, \mathbf{u}) + \varphi_1 f_1(\mathbf{x}, \mathbf{u}) + \varphi_2 f_2(\mathbf{x}, \mathbf{u}) + \cdots + \varphi_n f_n(\mathbf{x}, \mathbf{u}) \\ &= \varphi_0 f_0(\mathbf{x}, \mathbf{u}) + \boldsymbol{\varphi} \cdot \mathbf{f}(\mathbf{x}, \mathbf{u}) \end{aligned}$$

(H is called the *Hamiltonian*) and

$$M(\mathbf{x}, \boldsymbol{\varphi}) = \max_{\mathbf{v} \in \Omega} H(\mathbf{x}, \mathbf{v}, \boldsymbol{\varphi}).$$

Then the maximum principle can be stated.

Theorem 1 (Pontryagin [4]) Suppose $\mathbf{u}^*$ is an optimal control for the above problem and $\mathbf{x}^*$ is the corresponding trajectory. Then there exists a nonvanishing function $\boldsymbol{\varphi}^*(t) = (\varphi_1^*(t), \ldots, \varphi_n^*(t))$ and φ_0^* such that

(a) $$\dot{x}_i^* = \frac{\partial H}{\partial \varphi_i} = f_i(\mathbf{x}^*, \mathbf{u}^*), \qquad i = 1, 2, \ldots n,$$

(b) $$\dot{\varphi}_i^* = -\frac{\partial H}{\partial x_i} = -\varphi_0^* \frac{\partial f_0}{\partial x_i}(\mathbf{x}^*, \mathbf{u}^*) - \varphi_1^* \frac{\partial f_1}{\partial x_i}(\mathbf{x}^*, \mathbf{u}^*) - \cdots - \varphi_n^* \frac{\partial f_n}{\partial x_i^*}(\mathbf{x}^*, \mathbf{u}^*),$$

(c) φ_0^* is a nonpositive constant and

$$H(\mathbf{x}^*(t), \mathbf{u}^*(t), \boldsymbol{\varphi}^*(t)) = M(\mathbf{x}^*(t), \boldsymbol{\varphi}^*(t)) = \max_{\mathbf{v} \in \Omega} H(\mathbf{x}^*(t), \mathbf{v}, \boldsymbol{\varphi}^*(t)).$$

Further, $M(\mathbf{x}^*(t), \boldsymbol{\varphi}^*(t))$ is constant for $0 \le t \le t^*$.

Let us reapply this to the autonomous linear time optimal problem to check it.

We have

$$\dot{\mathbf{x}} = A\mathbf{x} + \mathbf{b}u = \mathbf{f}(\mathbf{x}, u), \qquad \Omega = [-1, 1],$$

and

$$C(u) = \int_0^t 1\, dt,$$

so

$$f_0 \equiv 1.$$

The Hamiltonian is just

$$H(\mathbf{x}, v, \boldsymbol{\varphi}) = \varphi_0 + \boldsymbol{\varphi}(A\mathbf{x} + \mathbf{b}v),$$

so the statement of the maximum principle becomes

$$H(\mathbf{x}^*(t), \mathbf{u}^*(t), \boldsymbol{\varphi}^*(t)) = \max_{|v| \le 1}(\varphi_0^* + \boldsymbol{\varphi}^*(t)(A\mathbf{x}(t) + \mathbf{b}v)),$$

which occurs simultaneously with the following maximum

$$\max_{|v| \le 1}(\boldsymbol{\varphi}^*(t)\mathbf{b}v).$$

This maximum clearly occurs when

$$u^*(t) = \operatorname{sgn}(\boldsymbol{\varphi}^*(t)\mathbf{b}),$$

and we calculate $\boldsymbol{\varphi}^*$ from (b) as just

$$\dot{\boldsymbol{\varphi}}^*(t) = -\boldsymbol{\varphi}^*(t)A,$$

which is just the adjoint equation. (Even in the general case (b) is called the adjoint equation.)

Remark If t_1 is fixed, and $\mathbf{x}_1$ is not given, then $\boldsymbol{\varphi}^*(t_1) = 0$. (This is a special case of a *transversality condition*, which we will investigate later in more detail.)

Consider a chemical reaction in which one component X is added at a constant rate over some fixed period of time $[0, T]$. Let $x(t)$ be the pH value of the reaction at time t. We assume that the quality of the final product depends critically on this pH value, and we control the pH by the strength u of some component of X. Further, suppose that the rate of change of pH is proportional to the sum of the current pH and the strength of the controlling ingredient u. That is,

$$\dot{x}(t) = \alpha x(t) + \beta u(t), \qquad x(0) = x_0, \qquad 0 \le t \le T,$$

where α and β are known positive constants and x_0 is the known initial pH.

We model the decrease in yield due to changes in the pH by

$$\int_0^T x^2 \, dt$$

and we suppose the rate of cost of maintaining the strength u to be proportional to u^2. The total cost associated with a control u will then be

$$C(u) = \int_0^T (ax^2 + u^2) \, dt,$$

where $a > 0$ is a given constant. In this example the control u is not constrained.

In our previous formulation,

$$\begin{aligned} f &= \alpha x + \beta u, \\ f_0 &= ax^2 + u^2, \\ H(x, u, \varphi) &= \varphi_0(ax^2 + u^2) + \varphi(\alpha x + \beta u). \end{aligned}$$

If the constant φ_0 is not zero, then by normalizing we can suppose it equals -1. The maximum principle allows $\varphi_0 = 0$; however, for such problems, the term in H involving the cost vanishes, and we would have the strange situation that our conditions for optimality do not depend on the cost of running the system! When $\varphi_0 = 0$ the system is called *abnormal*, and the maximum principle, in general, fails to determine u^*. In practice therefore, we suppose $\varphi_0 = -1$, and if we obtain nonsense from the maximum principle, keep in mind that we could have an abnormal system.

The equations given in (a) and (b) of the preceding theorem become

$$\dot{x}^* = \alpha x^* + \beta u^*, \qquad x^*(0) = x_0^*, \tag{1}$$

$$\dot{\varphi}^* = +2ax^* - \alpha\varphi^*, \qquad \varphi^*(T) = 0, \tag{2}$$

and the optimal control satisfies

$$H(x(t), u^*(t), \varphi^*(t)) = \max_v(-ax^*(t)^2 - v^2 + \varphi^*(t)(\alpha x^*(t) + \beta v)).$$

This maximum occurs at the same point as the maximum of

$$\max_v(-v^2 + \beta v\varphi^*(t)).$$

Since we are taking the maximum over all $v \in R$, for each t, and this maximum clearly exists, we can obtain it by differentiating with respect to v:

$$\frac{\partial}{\partial v}(-v^2 + \beta v\varphi^*(t)) = -2v + \beta\varphi^*(t) = 0;$$

hence the maximum occurs at $v = +\frac{1}{2}\beta\varphi^*(t)$, that is,

$$u^*(t) = +\tfrac{1}{2}\beta\varphi^*(t), \qquad 0 \le t \le T. \tag{3}$$

The optimal control is determined by (1)–(3), that is, we must solve the *two-point boundary-value problem*

$$\dot{x}^* = \alpha x^* + \frac{\beta^2}{2}\varphi^*, \qquad x^*(0) = x_0, \qquad 0 \le t \le T, \tag{4}$$

$$\dot{\varphi}^* = 2ax^* - \alpha\varphi^*, \qquad \varphi^*(T) = 0, \qquad 0 \le t \le T; \tag{5}$$

then $u^*(t) = \frac{1}{2}\beta\varphi^*(t)$, $0 \le t \le T$.

We can, in this case, manipulate these equations and find u^* as a function of time, an open-loop control. However, a more useful problem is to try to find a feedback control. In particular, can we find a function $C(t)$, $0 \le t \le T$, such that

$$u^*(t) = C(t)x^*(t), \qquad 0 \le t \le T?$$

In attempting to do this, divide (4) and (5) by x^*,

$$\frac{\dot{x}^*}{x^*} = \alpha + \frac{\beta^2}{2}\frac{\varphi^*}{x^*},$$

$$\frac{\dot{\varphi}^*}{x^*} = 2a - \alpha\frac{\varphi^*}{x^*},$$

and set $d(t) = \varphi^*(t)/x^*(t)$. Then

$$\dot{d} = -\left(\frac{\dot{x}^*\varphi^* - \dot{\varphi}^* x^*}{(x^*)^2}\right) = -\frac{\dot{x}^*}{x^*}\frac{\varphi^*}{x^*} + \frac{\dot{\varphi}^*}{x^*}$$

$$= -\left(\alpha + \frac{\beta^2}{2}d\right)d + 2a - \alpha d$$

or

$$\dot{d} = -\frac{\beta^2}{2}d^2 - 2\alpha d + 2a \qquad \text{with} \quad d(T) = 0. \tag{6}$$

To solve equation (6) (the Ricatti equation), make the change of variable

$$\frac{\dot{\xi}}{\xi} = \left(\frac{\beta^2}{2}\right)d, \qquad d = \frac{2}{\beta^2}\left(\frac{\dot{\xi}}{\xi}\right);$$

then (6) becomes

$$\frac{2}{\beta^2}\left(\frac{\xi\ddot{\xi} - (\dot{\xi})^2}{\xi^2}\right) = -\frac{2}{\beta^2}\left(\frac{\dot{\xi}}{\xi}\right)^2 - 2\alpha\left(\frac{2\dot{\xi}}{\beta^2\xi}\right) + 2a$$

$$+\frac{2}{\beta^2}\frac{\ddot{\xi}}{\xi} = -\frac{4\alpha}{\beta^2}\frac{\dot{\xi}}{\xi} + 2a,$$

that is,

$$\ddot{\xi} + 2\alpha\dot{\xi} - a\beta^2\xi = 0.$$

Since $d(T) = 0$, we have $\dot{\xi}(T) = 0$, and since we are only interested in the ratio $(\dot{\xi}/\xi)$, we can choose $\xi(T) = 1$. The solution is then

$$\xi(t) = e^{-\alpha(t-T)}\left[\cosh\sqrt{\alpha^2 + a\beta^2}(t - T) + \frac{\alpha}{\sqrt{\alpha^2 + a\beta^2}}\sinh\sqrt{\alpha^2 + a\beta^2}(t - T)\right]$$

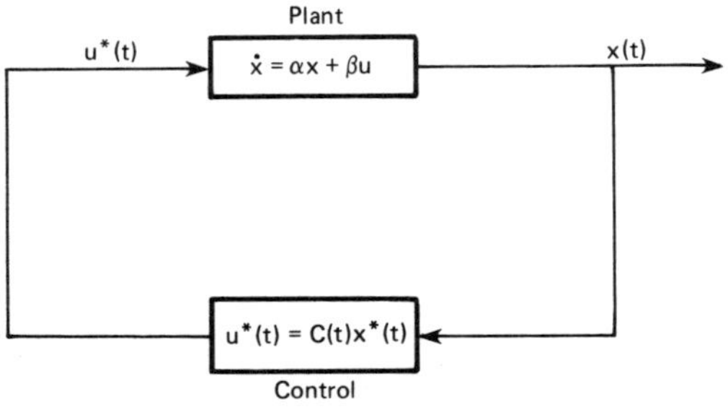

Fig. 1

for $0 \le t \le T$, and

$$u^*(t) = \frac{\beta}{2}\varphi^*(t) = \frac{\beta}{2}\frac{\varphi^*(t)}{x^*(t)}x^*(t) = \frac{\beta}{2}d(t)x^*(t)$$
$$= \frac{1}{\beta}\left(\frac{\dot{\xi}(t)}{\xi(t)}\right)x^*(t).$$

The function $\dot{\xi}/\xi$ can be precomputed, so we have a *linear feedback control* with *variable gain*

$$C(t) = \frac{1}{\beta}\frac{\dot{\xi}(t)}{\xi(t)}$$

(see Fig. 1).

2. CLASSICAL CALCULUS OF VARIATIONS

The oldest problem in the calculus of variations is probably the following:

Minimize the integral $\int_0^T f(y(x), y'(x))\,dx$ over all differentiable functions y passing through $y(0) = y_0$, $y(T) = y_1$, with y' piecewise continuous.

We can solve this problem with the maximum principle. To do this, we just reformulate it as a control problem.

Set $y' = u$, and since y' is not constrained, we take Δ, the set of admissible controls, to be all piecewise continuous functions. Then the above problem becomes a control problem with state equation

$$y' = u, \qquad y(0) = y_0, \qquad y(T) = y_1$$

and cost

$$C(u) = \int_0^T f(y, u)\, dt.$$

Applying the maximum principle, we set

$$H(y, u, \varphi) = -f(y, u) + \varphi u$$

and then, for the optimal u^*, y^*, φ^*, we have

$$\frac{dy^*}{dx} = u^*, \tag{1}$$

$$\frac{d\varphi^*}{dx} = -\frac{\partial H}{\partial y} = \frac{\partial f}{\partial y}(y^*, u^*), \tag{2}$$

and

$$H(y^*, u^*, \varphi^*) = \max_{v \in R}(-f(y, v) + \varphi v).$$

Since this function is differentiable with respect to v, a necessary test for its maximum is to set

$$\frac{\partial}{\partial v}(-f(y, v) + \varphi v) = 0, \tag{3}$$

and so

$$\varphi^* = \frac{\partial f}{\partial v}(y^*, u^*).$$

Combining (1)–(3) then gives us a partial differential equation for y^*,

$$\frac{d}{dx}\left(\frac{\partial f}{\partial y'}\right) - \frac{\partial f}{\partial y} = 0, \qquad y(0) = y_0, \qquad y(T) = y_1.$$

This last equation is known classically as the *Euler–Lagrange* equation [5].

As an example, suppose y_0 and y_1 are given points in the plane, and consider the problem of finding the *curve* of *shortest length* passing between y_0 and y_1 (see Fig. 2). If y is a curve passing between y_0 and y_1, then, parametrizing y with respect to x, $0 \leq x \leq 1$, say, its length is just

$$\int_0^1 \sqrt{1 + y'(x)}\, dx, \qquad y(0) = y_0, \qquad y(1) = y_1.$$

So our problem is

Minimize $\int_0^1 \sqrt{1 + (y')^2}\, dx$ subject to $y(0) = y_0$, $y(1) = y_1$.

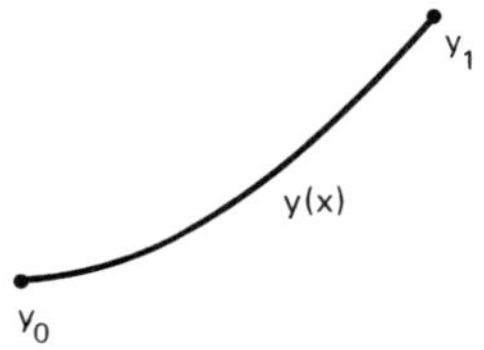

Fig. 2

For this problem $f(y, y') = \sqrt{1 + (y')^2}$, and the Euler equation becomes

$$\frac{d}{dx}\left(\frac{\partial f}{\partial y'}\right) - \frac{\partial f}{\partial y} = 0,$$

$$\frac{d}{dx}\left(\frac{y'}{\sqrt{1 + (y')^2}}\right) - 0 = 0,$$

that is,

$$\frac{y'}{\sqrt{1 + (y')^2}} = \text{const},$$

which has the solution $y' = \text{const}$, or y is linear.

Note that since $f = f(y, y')$ is not explicitly a function of x, Euler's equation can be rearranged in the form

$$f - y' f_{y'} = \text{const}.$$

This is called the *first integral of Euler's equation*, and it holds if f *is independent of* x.

Brachistochrone Problem

Find the curve passing between P_a and P_b, which has the property that (neglecting friction) a point mass would slide down the curve in minimum time (see Fig. 3).

This problem becomes: Find the curve y such that y' is continuous, $y(a) = P_a$, $y(b) = P_b$, and the integral

$$\frac{1}{\sqrt{2g}} \int_a^b \frac{\sqrt{1 + (y')^2}}{\sqrt{y}}\, dx$$

is minimized.

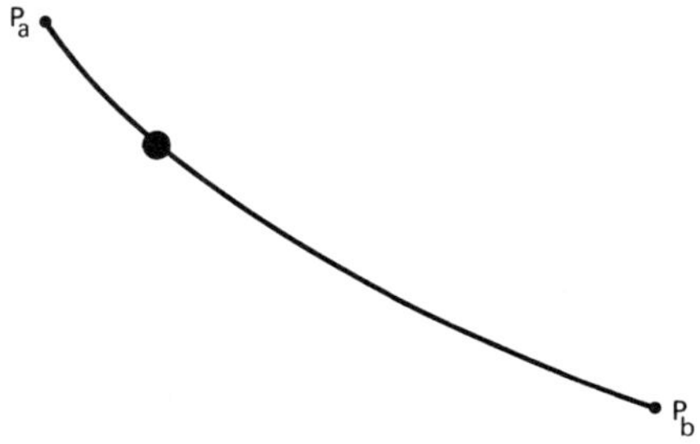

Fig. 3

For the equation of motion, notice that

$$\text{P.E.} + \text{K.E.} = \text{const.}$$

(where P.E. is potential energy and K.E. is kinetic energy), so

$$\tfrac{1}{2}mv^2 - mgy = E$$

where $v = ds/dt$ and $s =$ arc length. Initially $v = 0$, $y = 0$, hence $E = 0$, so at any time

$$\tfrac{1}{2}mv^2 = mgy,$$

$$\frac{ds}{dt} = \sqrt{2gy},$$

$$dt = \frac{ds}{\sqrt{2gy}} = \frac{\sqrt{1 + (dy/dx)^2}}{\sqrt{2gy}}\, dx.$$

In this problem

$$f(y, y') = \frac{\sqrt{1 + (y')^2}}{\sqrt{y}},$$

and so the first integral becomes

$$\frac{\sqrt{1 + (y')^2}}{\sqrt{y}} - (y')\frac{(1 + (y')^2)^{-1/2}}{\sqrt{y}}\, y' = \text{const},$$

that is,

$$\sqrt{y}\sqrt{1 + (y')^2} = A,$$

where A is a positive constant, or

$$y' = \sqrt{\frac{A - y}{y}}.$$

We try for a parametric solution of the form

$$y = A\sin^2(\tfrac{1}{2}\theta).$$

Then

$$A\sin\left(\frac{\theta}{2}\right)\cos\left(\frac{\theta}{2}\right)\frac{d\theta}{dx} = \frac{\cos(\frac{1}{2}\theta)}{\sin(\frac{1}{2}\theta)},$$

that is

$$dx = A\sin^2(\tfrac{1}{2}\theta)\,d\theta$$

or

$$x = B + \tfrac{1}{2}A(\theta - \sin\theta)$$

and

$$y = \tfrac{1}{2}A(1 - \cos\theta)$$

(these curves are called cycloids). We solve for A and B by finding the solution passing through $y = P_a$ when $x = a$, $y = P_b$ when $x = b$.

3. MORE EXAMPLES OF THE MAXIMUM PRINCIPLE

Example 1 Consider an inventory control model governed by the equations

$$\dot{I}(t) = P(t) - S(t), \qquad I(0) = I_0,$$
$$\dot{S}(t) = -\lambda P(t), \qquad S(0) = S_0,$$

where $I(t)$ is the level of inventory at time t, $S(t)$ the rate of sales at time t, $P(t)$ the rate of production with $\underline{P} \le P(t) \le \overline{P}$, and λ a positive constant. The first equation just expresses the assumption that the rate of change of the inventory level is the production rate minus the sales rate. The second reflects the assumption that the rate of change of sales declines at a rate proportional to the production rate.

Let c be the cost per unit time of producing at a rate $P(t)$, and h the cost per unit time of holding inventory level $I(t)$ (c, h constants). If T is given, we consider the problem of minimizing our total cost, that is,

$$\text{minimize } C(P) = \int_0^T (cP(t) + hI(t))\,dt$$

by choosing a suitable production rate P.

Applying the maximum principle we find

$$H = -cP - hI + \varphi_1(P - S) + \varphi_2(-\lambda P)$$

and

$$\dot{\varphi}_1 = -\frac{\partial H}{\partial I} = -(-h) = h,$$

$$\dot{\varphi}_2 = -\frac{\partial H}{\partial S} = -(-\varphi_1) = \varphi_1,$$

$\varphi_1(T) = \varphi_2(T) = 0$, as this is a free-end-point problem. Solving the adjoint equations gives

$$\varphi_1(t) = ht + c_1, \qquad \varphi_1(T) = 0 = hT + c_1, \qquad c_1 = -hT,$$

so

$$\varphi_1(t) = ht - hT,$$
$$\dot{\varphi}_2(t) = \varphi_1(t) = ht - hT,$$
$$\varphi_2(t) = \frac{ht^2}{2} - (hT)t + c_2;\ \varphi_2(T) = 0.$$

Therefore

$$c_2 = hT^2 - \frac{hT^2}{2} = \frac{hT^2}{2},$$

$$\varphi_2(t) = \frac{ht^2}{2} - (hT)t + \frac{hT^2}{2}.$$

Now, applying the last stage of the maximum principle, we see that

$$H(I^*(t), S^*(t), P^*(t), \varphi_1^*(t), \varphi_2^*(t)) = \max_{\underline{P} \le P \le \bar{P}} H(I^*(t), S^*(t), P, \varphi_1^*(t), \varphi_2^*(t)),$$

for all $t \in (0, T)$.

This maximum is achieved at the same time as the maximum of

$$\max_{\underline{P} \le P \le \bar{P}} (-cP + \varphi_1^*(t)P - \lambda\varphi_2^*(t)P) = \max_{\underline{P} \le P \le \bar{P}} P(-c + \varphi_1^*(t) - \lambda\varphi_2^*(t)).$$

That is, denoting

$$\xi(t) = -c + \varphi_1^*(t) - \lambda\varphi_2^*(t),$$

the optimal control for $0 \le t \le T$ is given by

$$P^*(t) = \begin{cases} \bar{P} & \text{when} \quad \xi(t) > 0 \\ \text{unknown} & \text{when} \quad \xi(t) = 0 \\ \underline{P} & \text{when} \quad \xi(t) < 0, \end{cases}$$

and from our remarks above,

$$\begin{aligned} \xi(t) &= -c + ht - hT - \lambda\left(\frac{ht^2}{2} - (hT)t + \frac{hT^2}{2}\right) \\ &= -\frac{\lambda h t^2}{2} + (h + \lambda h T)t - \frac{\lambda h T^2}{2} - hT - c. \end{aligned}$$

Note that $\xi(t)$ is a quadratic, so we can have at most two switches. Solving for the zeros of ξ, we find that

$$t_{\pm} = \frac{1 + \lambda T}{\lambda} \pm \frac{1}{\lambda}\sqrt{1 - \frac{2c\lambda}{h}}$$

are the switching times. Note that

$$t_{\pm} = T + \frac{1}{\lambda}\left(1 \pm \sqrt{1 - \frac{2c\lambda}{h}}\right)$$

and so with $\lambda, c, h > 0$, the only optimal strategy is

$$P^*(t) = \underline{P}, \qquad 0 \le t \le T.$$

Example 2 (Isaacs [3]) We return to Example 3 of Chapter 1. For simplicity, we assume that the constant $k = 1$ and

$$\dot{x} = ux, \qquad x(0) = x_0,$$

$$\max \int_0^T (1 - u(t))x(t)\,dt.$$

The Hamiltonian is

$$H(x, u, \varphi) = (u - 1)x + \varphi u x,$$

and the adjoint equation is

$$\dot{\varphi} = -\frac{\partial H}{\partial x} = -(u - 1 + \varphi u) = 1 - u(\varphi + 1), \qquad \varphi(T) = 0.$$

We then have to minimize over $0 \le u \le 1$,

$$u(x + \varphi x) = ux(1 + \varphi),$$

and consequently the optimum control is given by

$$u^*(t) = \begin{cases} +1 & \text{if } 1 + \varphi < 0 \\ 0 & \text{if } 1 + \varphi > 0 \end{cases} \qquad (x \text{ is always positive}). \qquad (1)$$

Since $\varphi(T) = 0$, $u^*(T) = 0$, and so the optimum control must be zero over the final time period. Using this information, we can solve the adjoint equation backward from $t = T$, with $\varphi(T) = 0$ and $u^* = 0$. Then

$$\varphi(t) = t - T \qquad \text{for} \quad t \leq T.$$

The control will switch when $1 + \varphi(t) = 0$, that is, when

$$1 + t - T = 0 \qquad \text{or} \quad t = T - 1,$$

and so

$$\varphi(t) = t - T, \qquad u^*(t) = 0, \qquad \text{for} \quad t \in (T - 1, T).$$

Since φ satisfies the adjoint equation, it is continuous over $(0, T)$; so, over the penultimate switching interval

$$\dot{\varphi} = -\varphi, \qquad \varphi(T - 1) = -1,$$

that is,

$$\varphi(t) = -e^{T-t-1}, \qquad t \leq T - 1.$$

However, the exponential function can equal -1 at most once, so we must have

$$\begin{aligned} \varphi(t) &= -e^{T-t-1}, & 0 \leq t \leq T - 1, \\ u^*(t) &= +1, & 0 \leq t \leq T - 1. \end{aligned}$$

The optimal policy (if it exists) must be full investment up to one time period before the final time, and then flat-out consumption.

The cost of u^* is obtained from

$$\dot{x}^* = x^*, \qquad x^*(0) = x_0 \qquad \text{over} \quad [0, T - 1],$$

that is,

$$x^*(t) = x_0 e^t, \qquad 0 \leq t \leq T - 1.$$

Then over $T - 1 \leq t \leq T$, $\dot{x}^* = 0$, that is, $x^*(t) = x_0 e^{T-1}$ since x^* is continuous. Hence

$$C(u^*) = \int_0^T (1 - u^*)x^* \, dt = \int_{T-1}^T (1 - u^*)x^*(t) \, dt = x_0 e^{T-1}.$$

Example 3 An Inventory Control Problem (Bensoussan *et al.* [1]) Consider an inventory of items being controlled over a time period $[0, T]$. If $d(t)$ represents the demand rate for these items at time t, and $u(t)$ the production rate at time t, then the rate of change of the inventory level $x(t)$ will be just

$$\dot{x}(t) = u(t) - d(t), \qquad x(0) = x_0, \qquad 0 \le t \le T. \tag{2}$$

Suppose the firm chooses desired levels $u_d(t)$, $x_d(t)$, $0 \le t \le T$, for the production and inventory. In order to penalize variations in u and x from the desired levels, we could formulate the cost of a production rate u as

$$C(u) = \int_0^T \{c(u(t) - u_d(t))^2 + h(x(t) - x_d(t))^2\}\, dt \tag{3}$$

for some constants $c, h > 0$. Then the control problem would be to minimize $C(u)$ subject to (2). The problem (2), (3) would appear more realistic if we placed constraints on u and x, such as $x(t) \ge 0$, $u(t) \ge 0$, $t > 0$. For the moment, we shall assume that u_d and x_d are sufficiently large that the minimization of the cost (3) keeps us in the region $x \ge 0$, $u \ge 0$. In Chapter 9 we shall consider inventory problems with state and control constraints.

For this problem, the Hamiltonian is

$$H(x, u, \psi) = -c(u - u_d)^2 - h(x - x_d)^2 + \psi(u - d),$$

and the adjoint equation is

$$\dot{\psi}^* = -\frac{\partial H}{\partial x} = 2h(x^* - x_d), \qquad \psi^*(T) = 0, \tag{4}$$

since (2), (3) is a free-end-point control problem. The maximum of the Hamiltonian over all $u \in R$ will occur at

$$\frac{\partial H}{\partial u} = -2c(u - u_d) + \psi = 0,$$

that is,

$$u^*(t) = \frac{\psi^*(t)}{2c} + u_d(t), \qquad 0 \le t \le T. \tag{5}$$

Substituting (5) into (2) gives

$$\dot{x}^*(t) = \frac{\psi^*(t)}{2c} + u_d(t) - d(t), \qquad 0 \le t \le T, \qquad x^*(0) = x_0, \tag{6}$$

which, together with (4) forms a two-point boundary-value problem for the optimum trajectory and adjoint variable.

Rather than solve this problem directly, we attempt a solution for the adjoint variable in the form

$$\psi^*(t) = a(t) + b(t)x^*(t), \qquad t \geq 0, \tag{7}$$

for some functions a and b. The advantage is again that once the functions a and b are determined, relation (7) and (5) gives the optimal control in feedback form. Differentiating (7) gives

$$\dot{\psi}^* = \dot{a} + \dot{b}x^* + b\dot{x}^*,$$

and substituting (6) and (4) for $\dot{x}, \dot{\psi}$ results in

$$2h(x^* - x_{\mathrm{d}}) = \dot{a} + \dot{b}x^* + b\left(\frac{\psi^*}{2c} + u_{\mathrm{d}} - d\right).$$

Plugging in (7) for ψ^* results in

$$\dot{a}(t) + b(t)(u_{\mathrm{d}}(t) - d(t)) + 2hx_{\mathrm{d}}(t)$$
$$+ \frac{a(t)b(t)}{2c} + \left(\dot{b}(t) + \frac{b^2(t)}{2c} - 2h\right)x^*(t) = 0 \quad \text{for all } 0 \leq t \leq T. \tag{8}$$

This will be satisfied if we choose a and b such that

$$\dot{b}(t) + \frac{b^2(t)}{2c} - 2h = 0, \tag{9}$$

$$\dot{a}(t) + b(t)(u_{\mathrm{d}}(t) - d(t)) + 2hx_{\mathrm{d}}(t) + \frac{a(t)b(t)}{2c} = 0. \tag{10}$$

From $\psi^*(T) = 0$, without loss of generality we can also suppose

$$b(T) = 0, \tag{11}$$

$$a(T) = 0. \tag{12}$$

The equation for b is a Ricatti equation, which can be solved by the substitution $b(t) = \dot{\xi}(t)/\xi(t)$ to give

$$b(t) = -2c\sqrt{\frac{h}{c}}\tanh\left(\sqrt{\frac{h}{c}}(T-t)\right), \qquad 0 \leq t \leq T. \tag{13}$$

When (13) is substituted in (10) we are left with a linear equation for a, which can be solved by variation of parameters.

For simplicity, we take a particular case $u_d(t) = d(t)$, $x_d(t) = c_d$, a constant, for $0 \le t \le T$. That is, the firm wishes to have the production rate match the demand rate while maintaining a constant level of inventory. Now a satisfies

$$\dot{a}(t) + \frac{a(t)b(t)}{2} + 2hc_d = 0, \qquad a(T) = 0,$$

which, when solved, gives

$$a(t) = 2c\left(\sqrt{\frac{h}{c}}\right)c_d \tanh\left(\sqrt{\frac{h}{c}}(T-t)\right). \tag{14}$$

substitution of (14) and (13) into (5) gives the feedback control law

$$u^*(t,x) = \sqrt{\frac{h}{c}}\,[c_d - x]\tanh\left(\sqrt{\frac{h}{c}}(T-t)\right) + d(t), \qquad 0 \le t \le T. \tag{15}$$

As a consequence the optimum control rate is equal to the demand rate plus an inventory correction factor, which tends to restore the inventory to the desired level c_d.

Further computation gives the optimal inventory level as

$$x^*(t) = c_d + \frac{(x_0 - c_d)}{\cosh(\sqrt{h/c}\,T)}\cosh\left[\sqrt{\frac{h}{c}}(T-t)\right], \qquad 0 \le t \le T. \tag{16}$$

We can see from (15) and (16) that if we start out with $x_0 = c_d$, the desired inventory, we remain at that level and meet the demand through production. If we start out away from c_d, the inventory level asymptotically moves toward the desired level.

Example 4 Moon-Landing Problem (Fleming and Rishel [2]) This problem was described in Chapter 1. The state equations were

$$\begin{aligned} \dot{h} &= v, \\ \dot{v} &= -g + u/m \qquad \text{with cost} \quad C(u) = \min_u \int_0^T u(\xi)\,d\xi, \\ \dot{m} &= -ku, \end{aligned} \tag{17}$$

where

$$\begin{gathered} 0 \le u(t) \le 1, \\ (h(0), v(0), m(0)) = (h_0, v_0, M + F), \\ h(T) = 0, \qquad v(T) = 0, \quad m(T) > 0. \end{gathered}$$

The Hamiltonian is

$$H(h, v, m, u, \psi_1, \psi_2, \psi_3) = -u + \psi_1 v + \psi_2(-g + u/m) - k\psi_3 u.$$

The adjoint equations become

$$\begin{aligned} \dot{\psi}_1 &= -\frac{\partial H}{\partial h} = 0, \\ \dot{\psi}_2 &= -\frac{\partial H}{\partial v} = -\psi_1, \\ \dot{\psi}_3 &= -\frac{\partial H}{\partial m} = \frac{\psi_2 u}{m^2}, \end{aligned} \tag{18}$$

with $\psi_3(T) = 0$ since $m(T)$ is not specified.

The maximum with respect to u of the Hamiltonian will occur at the same point as

$$\max_{0 \le u \le 1} [-u + \psi_2 u/m - k\psi_3 u],$$

that is,

$$u^*(t) = \begin{cases} 1 & \text{when} \quad 1 - \psi_2(t)/m + k\psi_3(t) < 0 \\ \text{undefined} & \text{when} \quad 1 - \psi_2(t)/m + k\psi_3(t) = 0 \\ 0 & \text{when} \quad 1 - \psi_2(t)/m + k\psi_3(t) > 0. \end{cases}$$

Note that for the problem to be physically reasonable,

$$\text{max thrust} > \text{gravitational force},$$

that is,

$$1 > (M + F)g \qquad \text{or} \qquad \frac{1}{M + F} > g.$$

In general, if max thrust $= \alpha$ $(0 \le u(t) \le \alpha)$, then

$$\frac{\alpha}{M + F} > g.$$

Our intuition would lead us to expect that the optimal control is first a period of free fall ($u^* \equiv 0$) followed by maximum thrust ($u^* \equiv 1$), culminating, we hope, in a soft landing. Assuming this to be the case, we shall first show how to construct such an optimal control and then use this information to show that it is the unique solution of the maximum principle.

Suppose that $u^*(t) = +1$ over the last part of the trajectory $[\xi, T]$. Remember that $h(T) = 0$, $v(T) = 0$, $m(T)$ is unknown, and $m(\xi) = M + F$ since we have coasted to this point. Then the solution of (17) is

$$h(\xi) = -\frac{1}{2}g(T-\xi)^2 - \frac{M+F}{k^2}\ln\left(\frac{M+F-k(T-\xi)}{M+F}\right) - \frac{T-\xi}{k},$$
$$v(\xi) = g(T-\xi) + \frac{1}{k}\ln\left(\frac{M+F-k(T-\xi)}{M+F}\right), \tag{19}$$
$$m(\xi) = M + F.$$

If we plot $h(\xi)$ against $v(\xi)$, we get the curve shown in Fig. 4.

Clearly this curve is the *locus* of all (height, velocity) pairs that we can *steer* to (0, 0) with *full thrust* +1.

There are some physical restrictions on the length of this curve. Namely, as the spacecraft is burning fuel at a rate k, the total amount of fuel will be burned in time F/k seconds.

Consequently,

$$0 \leq T - \xi \leq F/k.$$

Over the first part of the trajectory $[0, \xi]$, we "free-fall," $u^* = 0$, and we have

$$h(t) = -\tfrac{1}{2}gt^2 + v_0 t + h_0,$$
$$v(t) = -gt + v_0, \tag{20}$$
$$m(t) = M + F,$$

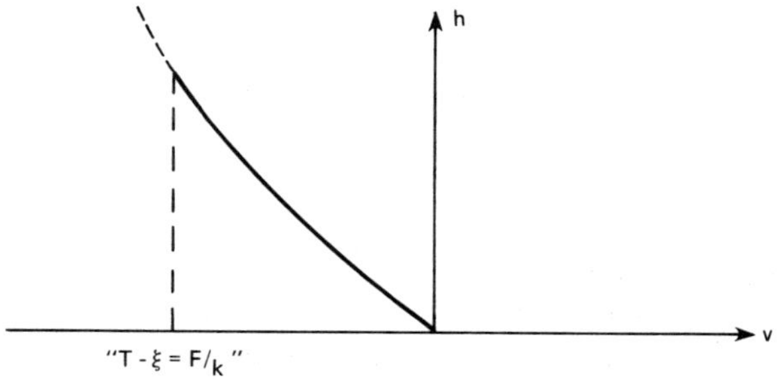

Fig. 4

or, in the phase plane,

$$h(t) = h_0 - \frac{1}{2g}[v^2(t) - v_0^2], \qquad 0 \le t \le \xi. \tag{21}$$

As expected, this curve is a parabola.

Using this strategy we free-fall following trajectory (21) until we hit the curve constructed previously in Fig. 4., where we switch to full thrust (see Fig. 5). The switching time ξ is the time at which (19) and (20) cross.

We now show that this choice for u^* satisfies the maximum principle. Let

$$\psi_1(0) = \lambda_1, \qquad \psi_2(0) = \lambda_2, \qquad \psi_3(0) = \lambda_3.$$

From the adjoint equations (18)

$$\begin{aligned} \psi_2(t) &= \lambda_2 - \lambda_1 t, && 0 \le t \le T, \\ \psi_3(t) &= \lambda_3, && 0 \le t \le \xi, \\ m(t) &= k(\xi - t) + M + F, && \xi \le t \le T, \end{aligned}$$

we see that

$$\psi_3(\tau) = \lambda_3 + \int_\xi^\tau \frac{(\lambda_2 - \lambda_1 t)}{[k(\xi - t) + M + F]^2}\,dt \qquad \xi \le \tau \le T.$$

Since the switching function

$$r(t) = 1 - \frac{\psi_2(t)}{m} + k\psi_3(t), \qquad 0 \le t \le T,$$

is zero at $t = \xi$,

$$r(\xi) = 1 - \frac{\lambda_2 - \lambda_1 \xi}{M + F} + k\lambda_3 = 0, \tag{22}$$

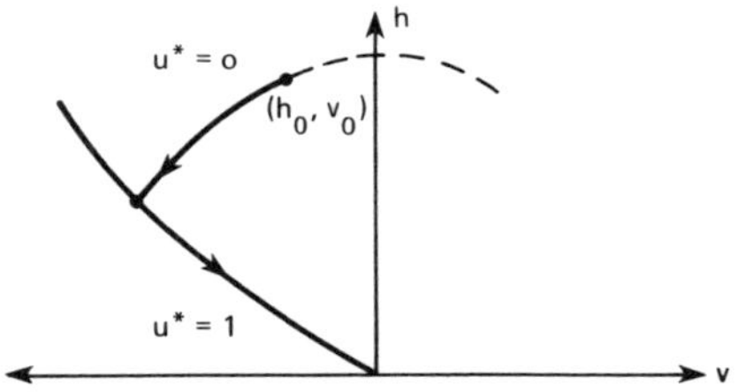

Fig. 5

and

$$\psi_3(T) = 0, \tag{23}$$

it is not too hard to find three unknowns $\lambda_1, \lambda_2, \lambda_3$ satisfying (22) and (23). They can also be chosen so that $r(t) < 0$ on $[0, \xi]$, and $r(t) > 0$ on $[\xi, T]$. Any such solution clearly satisfies the maximum principle.

We shall next show that this is the only type of solution of the maximum principle. Note that

$$\dot{r}(t) = -\lambda_1/m(t), \qquad 0 \le t \le T,$$

and so if $\lambda_1 \neq 0$, $r(t)$ is strictly monotone, and the optimal control is either of the type above or the opposite, namely, $u^* = 1$, then $u^* = 0$. We now show that if (h_0, v_0) lies above the switching curve a control $u^* = 1$, then $u^* = 0$ cannot achieve a soft landing. If there were such a control, with switch at τ, say, then $v(T) = 0$ only if

$$T = v(\tau)/g + \tau \qquad \text{(by (17))}.$$

Now since $h(T) = 0$, we must have by (17)

$$h(\tau) = -v(\tau)^2/2g < 0,$$

and we switch below the surface of the moon, which is not too good for the astronaut.

Last, notice that if (h_0, v_0) is below the switching curve, a soft landing is not possible, because even by using full thrust for the entire trajectory, the spacecraft will strike the moon with a nonzero velocity (Fig 5).

In the case $\lambda_1 = 0$, $r(t)$ is a constant and if nonzero, the optimal control must be $u^* \equiv 1$ or $u^* \equiv 0$; the last case can easily be seen to be false, as is the former, unless (h_0, v_0) lies on the curve (19). The only remaining case is then $\lambda_1 = 0$ and $r(t) \equiv 0$, $0 \le t \le T$.

First, $\lambda_1 = 0$ implies $\psi_1(t) \equiv 0$ and so $\psi_2(t) \equiv \lambda_2$, $0 \le t \le T$. So $r(t) \equiv 0$, $0 \le t \le T$, means that

$$1 - \frac{\lambda_2}{m(t)} + k\psi_3(t) \equiv 0 \qquad \text{for} \quad 0 \le t \le T,$$

which implies that the functions $\{1, 1/m(t), \psi_3(t)\}$ are linearly dependent, a contradiction.

Since we have eliminated all other cases, the given solution is the only solution of the maximum principle.

We can simplify the equation of the switching curve somewhat by approximating

$$\ln\left(1 - \frac{k(T-\xi)}{M+F}\right) \simeq -\frac{k(T-\xi)}{M+F} - \frac{k^2(T-\xi)^2}{(M+F)^2}.$$

Then

$$h(\xi) \sim a(T-\xi)^2,$$
$$v(\xi) \sim -2a(T-\xi) - b(T-\xi)^2,$$

where

$$a = \frac{1}{2}\left(\frac{k - g(M+F)}{M+F}\right),$$

$$b = \frac{k}{2(M+F)^2}.$$

Now the switching curve can be approximately represented by

$$\Psi(h, v) = \frac{b}{a}h + 2a\sqrt{\frac{h}{a}} + v = 0,$$

so if we are measuring h and v simultaneously, we free-fall as long as $\Psi(h, v) > 0$, and switch to maximum thrust when first $\Psi(h, v) = 0$.

Problems

1. Find the production rate $P(t)$ that will change the inventory level $I(t)$ from $I(0) = 2$ to $I(1) = 1$ and the sales rate $S(t)$ from $S(0) = 0$ to $S(1) = 1$ in such a way that the cost $C = \int_0^1 P^2(t)\,dt$ is a minimum. Assume

$$\dot{I}(t) = P(t) - S(t),$$
$$\dot{S}(t) = -P(t)$$

(P is unrestricted).

2. Minimize $C(u) = \frac{1}{4}\int_0^1 u(t)^4\,dt$ subject to

$$\dot{x} = x + u, \qquad x(0) = x_0, \qquad x(1) = 0.$$

3. Find the extremals for

(a) $\int_0^\pi ((y')^2 - y^2)\,dx, \qquad y(0) = 0, \qquad y(\pi) = 0,$

(b) $\int_0^1 ((y')^2 + 4xy')\,dx, \qquad y(0) = 0, \qquad y(1) = 1.$

Remember that the extremals for

$$\int_a^b f(x, y'(x), y(x))\,dx, \qquad y(a) = y_a, \qquad y(b) = y_b$$

are the solutions of the Euler–Lagrange equation

$$f_y - \frac{d}{dx}(f_{y'}) = 0$$

that satisfy the boundary conditions.

4. Consider the following model for determining advertising expenditures for a firm that produces a single product. If $S(t)$ is the sales rate at time t and $A(t)$ the advertising rate at time t, restricted so that $0 \le A(t) \le \bar{A}$, then we assume

$$S(t) = -\lambda S(t) + \gamma \int_0^t A(t-\tau)e^{-\tau}\,d\tau, \qquad \lambda, \gamma > 0.$$

We wish to choose A to maximize $\int_0^T S(t)\,dt$, the total sales over period $[0, T]$.

5. Find the optimal control which maximizes the cost

$$\int_0^{100} x(t)\,dt$$

subject to the system dynamics

$$\dot{x}(t) = -0.1x(t) + u(t), \qquad 0 \le u(t) \le 1, \qquad x(0) = x_0.$$

6. Apply the maximum principle to the control problem

$$\text{maximize} \int_0^2 (2x(t) - 3u(t) - \alpha u^2(t))\,dt, \qquad \alpha \ge 0$$

$$\text{subject to} \quad \dot{x}(t) = x(t) + u(t), \qquad x(0) = 5, \qquad 0 \le u(t) \le 2.$$

7. Discuss the optimal control of the system

$$\dot{x}_1 = x_2 + u, \qquad \dot{x}_2 = -u$$

from $x(0) = (x_1^0, x_2^0)$ to $(0, 0)$, which minimizes the cost

$$\int_0^t x_1^2(t)\,dt.$$

8. What is the solution of the maximum principle for the problem of minimum fuel consumption

$$\text{minimize} \int_0^{t_1} \sqrt{1 + u(t)^2}\,dt,$$

for the system

$$\ddot{x} = u, \qquad |u| \leq 1,$$

$$x(0) = x_1^0, \qquad \dot{x}(0) = x_2^0, \qquad x(t_1) = 0, \qquad \dot{x}(t_1) = 0.$$

Does the solution of this problem change if the cost is replaced by

$$\int_0^{t_1} |u(t)|\, dt \,?$$

9. A function $E(x^*, u^*, u, t)$ is defined by

$$E(x^*, u^*, u, t) = f(x^*, u, t) - f(x^*, u^*, t) + (u^* - u) f_u(x^*, u^*, t).$$

Show that if (x^*, u^*) is an optimal solution of the calculus of variations problem of Section 2, then

$$E(x^*, u^*, u, t) \geq 0 \qquad \text{for all} \quad u, t.$$

(This is the *Weierstrass E function* necessary condition for a strong extremum.)

References

[1] A. Bensoussan, E. Hurst, and B. Naslund, "Management Applications of Modern Control Theory." North-Holland Publ., New York, 1974.

[2] W. Fleming and R. Rishel. "Deterministic and Stochastic Optimal Control." Springer-Verlag, Berlin and New York, 1975.

[3] R. Isaacs, "Differential Games." Wiley, New York, 1965.

[4] L. Pontryagin, V. Boltyanskii, R. Gramkrelidze, and E. Mischenko, "The Mathematical Theory of Optimal Processes." Wiley (Interscience), New York, 1962.

[5] H. Sagan, "Introduction to the Calculus of Variations." McGraw-Hill, New York, 1969.

Chapter IV The General Maximum Principle; Control Problems with Terminal Payoff

1. INTRODUCTION

In this chapter we consider the maximum principle for the general nonautonomous control problem. Namely, suppose that

(1) $\dot{\mathbf{x}} = \mathbf{f}(\mathbf{x}, t\,\mathbf{u})$ is the control process and $\mathbf{f}$ is continuously differentiable in R^{n+1+m};

(2) we are given sets X_0, X_1 in R^n to be interpreted as sets of allowable initial and final values, respectively;

(3) $\Delta = \{\mathbf{u} : \mathbf{u}$ is bounded, piecewise continuous $\mathbf{u}(t) \in \Omega \subset R^m$, and $\mathbf{u}$ steers some initial point $\mathbf{x}_0 \in X_0$, at a fixed time t_0, to some final point $\mathbf{x}_1 \in X_1$, at time $t_1\}$;

(4) The cost of control $\mathbf{u}$ is

$$C(\mathbf{u}) = \int_{t_0}^{t_1} f_0(\mathbf{x}(t), t, \mathbf{u}(t))\, dt,$$

f_0 continuously differentiable in R^{n+m+1}.

Define the Hamiltonian for $\mathbf{x} = (x_1, \ldots, x_n)$, $\mathbf{u} = (u_1, \ldots, u_m)$, and $\boldsymbol{\psi} = (\psi_0, \psi_1, \ldots, \psi_n)$ as

$$H(\mathbf{x}, \mathbf{u}, t, \boldsymbol{\psi}) = \psi_0 f^0(\mathbf{x}, t, \mathbf{u}) + \psi_1 f_1(\mathbf{x}, t, \mathbf{u}) + \cdots + \psi_n f_n(\mathbf{x}, t, \mathbf{u})$$

and define

$$M(\mathbf{x}, t, \boldsymbol{\psi}) = \max_{\mathbf{u} \in \Omega} H(\mathbf{x}, \mathbf{u}, t, \boldsymbol{\psi}).$$

Theorem 1 (Lee and Marcus [2]) If $\mathbf{u}^*$, $\mathbf{x}^*$ are optimal for the above problem over $[t_0, t_1^*]$, then there exists a nontrivial adjoint response $\boldsymbol{\psi}^*$ such that

(i) $\dot{\mathbf{x}}^*(t) = \mathbf{f}(\mathbf{x}^*(t), t, \mathbf{u}^*(t))$,

(ii) $\dot{\psi}_j^*(t) = -\dfrac{\partial H}{\partial x_j} = \sum_{i=0}^{n} \psi_i^*(t) \dfrac{\partial f_i}{\partial x_j}(\mathbf{x}^*(t), t, \mathbf{u}^*(t)), \qquad j = 1, \ldots, n,$

(iii) ψ_0^* is a nonpositive constant (≤ 0), and

(iv) $H(\mathbf{x}^*(t), \mathbf{u}^*(t), t, \boldsymbol{\psi}^*(t)) = M(\mathbf{x}^*(t), t, \boldsymbol{\psi}^*(t))$, at points of continuity. Further

$$M(\mathbf{x}^*(t), t, \boldsymbol{\psi}^*(t)) = \int_{t_1^*}^{t} \sum_{i=0}^{n} \psi_i^*(s) \frac{\partial f_i}{\partial t}(\mathbf{x}^*(s), s, \mathbf{u}^*(s))\, ds$$

and, hence

$$M(\mathbf{x}^*(t_1^*), t_1^*, \boldsymbol{\psi}^*(t_1^*)) = 0.$$

If X_0 and X_1 (or just one of them) are manifolds in R^n with tangent spaces T_0 and T_1 at $\mathbf{x}^*(t_0)$ and $x^*(t_1^*)$, respectively, then $\boldsymbol{\psi}^*$ can be selected to satisfy the *transversality* conditions (or at just one end)

$$\boldsymbol{\psi}^*(t_0) \perp T_0 \qquad \text{and} \qquad \boldsymbol{\psi}^*(t_1^*) \perp T_1.$$

In the most usual case X_1 (and possibly X_0) will be of the form

$$X_1 = \{\mathbf{x} : g_k(\mathbf{x}) = 0,\ k = 1, 2, \ldots, l\}. \tag{1}$$

for some given real-valued functions $\{g_1, \ldots, g_l\}$. For example, our fixed-end-point problem could be formulated $\mathbf{x}^*(t_1^*) = \mathbf{x}^1$ or

$$g_r(\mathbf{x}^*(t_1)) = x_r^*(t_1) - x_r^1 = 0, \qquad r = 1, \ldots, n. \tag{2}$$

If the functions $\{g_1, \ldots, g_l\}$ are differentiable, then the transversality condition can be simply written

$$\boldsymbol{\psi}^*(t_1^*) = \sum_{k=1}^{l} \lambda_k \nabla g_k(\mathbf{x}^*(t_1^*)),$$

for some constants $\{\lambda_1, \ldots, \lambda_l\}$. By (1) at the terminal point we have also

$$g_k(\mathbf{x}^*(t_1^*)) = 0 \qquad \text{for} \quad k = 1, \ldots, l.$$

For example, in the fixed-end-point problem,

$$\nabla g_r(\mathbf{x}^*(t_1^*)) = (0, \ldots, \underset{r\text{th}}{1}, 0 \ldots 0).$$

So (3) becomes $\psi^*(t_1^*) = (\lambda_1, \lambda_2, \ldots, \lambda_n)$, and in this case (as we have been assuming all along) $\psi^*(t_1)$ is unknown.

In the other case that we have been using, the *free-end-point problem*, $X_1 = R^n$; consequently the only vector orthogonal to the tangent space is $\mathbf{0}$, that is,

$$\psi^*(t_1^*) = \mathbf{0}.$$

2. CONTROL PROBLEMS WITH TERMINAL PAYOFF

In many applications of control theory the cost takes the form

$$C(\mathbf{u}) = g(\mathbf{x}(T)) + \int_0^T f_0(\mathbf{x}(t), t, \mathbf{u}(t))\,dt, \tag{1}$$

where $T > 0$ is fixed and $g: R^n \to R$ is a given continuously differentiable function that represents some terminal payoff, or salvage cost, at the final time T. We can convert such problems into the standard form by adding an extra state variable $x_{n+1}(t)$ satisfying

$$\dot{x}_{n+1}(t) = 0, \qquad x_{n+1}(T) = g(\mathbf{x}(T))/T.$$

Then $g(x(T)) = \int_0^T x_{n+1}(t)\,dt$, and the cost (1) is now in purely integral form. When the maximum principle is applied to this new problem, one finds the only change is in the transversality conditions, which now become

$$\psi_i^*(T) = -\frac{\partial g}{\partial x_i}(\mathbf{x}^*(T)), \qquad i = 1, 2, \ldots, n. \tag{2}$$

Example 1 A Continuous Wheat-Trading Model (Norström [3]) We consider the following model for a firm engaged in buying and selling wheat, or a similar commodity. The firm's assets are of two types, cash and wheat, and we represent the balance of each quantity at time t by $x_1(t)$ and $x_2(t)$, respectively. The initial assets $x_1(0)$ and $x_2(0)$ are given. The price of wheat over the planning period $[0, T]$ is assumed to be known in advance and is denoted it by $p(t)$, $0 \le t \le T$.

The firm's objective is to buy and sell wheat over the period $[0, T]$ so as to maximize the value of its assets at time T, that is, it wishes to maximize

$$C = x_1(T) + p(T)x_2(T). \tag{3}$$

If the rate of buying or selling wheat at time t is denoted by $u(t)$, ($u(t) > 0$ indicates buying; $u(t) < 0$, selling), then we could model the operation of the firm by the differential equations

$$\dot{x}_1(t) = -\alpha x_2(t) - p(t)u(t), \tag{4}$$

$$\dot{x}_2(t) = u(t), \tag{5}$$

where $\alpha > 0$ in (4) is the cost associated with storing a unit of wheat, and the term $p(t)u(t)$ indicates the cost (or revenue) of purchases (or sales) of wheat at time t. We have also the natural control constraint

$$\underline{M} \leq u(t) \leq \bar{M}, \qquad \underline{M}, \bar{M} \quad \text{given constants.} \tag{6}$$

The Hamiltonian for this problem is

$$H = \psi_1(-\alpha x_2 - pu) + \psi_2 u, \tag{7}$$

and the adjoint equations are

$$\dot{\psi}_1 = -\frac{\partial H}{\partial x_1} = 0, \tag{8}$$

$$\dot{\psi}_2 = -\frac{\partial H}{\partial x_2} = \alpha\psi_1 \tag{9}$$

with transversality conditions

$$\psi_1(T) = -1, \qquad \psi_2(T) = -p(T)$$

by (2). In this case (8) and (9) are independent of the state variables and can be solved directly to give

$$\psi_1(t) = -1, \tag{10}$$

$$\psi_2(t) = -\alpha(t - T) - p(T), \qquad 0 \leq t \leq T. \tag{11}$$

This enables us to write out the Hamiltonian explicitly as

$$\begin{aligned} H &= \alpha x_2 + p(t)u + u(-\alpha(t - T) - p(T)) \\ &= u(p(t) - \alpha(t - T) - p(T)) + \alpha x_2. \end{aligned}$$

H will be minimized as a function of u if

$$u^*(t) = \begin{cases} M \text{ (buy)} & \text{when} \quad p(t) < p(T) - \alpha(T - t)(= -\psi_2(t)) \\ \underline{M} \text{ (sell)} & \text{when} \quad p(t) > p(T) - \alpha(T - t) \\ \text{undetermined} & \text{when} \quad p(t) = p(T) - \alpha(T - t). \end{cases} \tag{12}$$

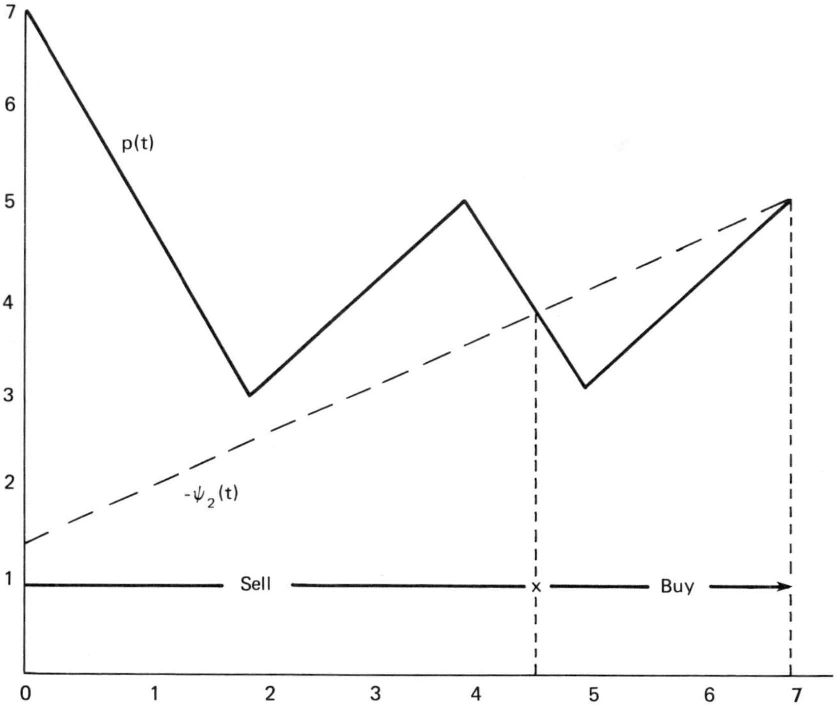

Fig. 1

Figure 1 illustrates this solution for a particular price function,

$$T = 7, \quad \alpha = \tfrac{1}{2}, \quad \underline{M} = -1, \quad \bar{M} = 1, \quad x_1(0) = 50, \quad x_2(0) = 1,$$

$$p(t) = \begin{cases} -2t + 7 & 0 \le t \le 2 \\ t + 1 & 2 \le t \le 4 \\ -2t + 13 & 4 \le t \le 5 \\ t - 2 & 5 \le t \le 7. \end{cases}$$

From (11), it follows that $\psi_2(t) = -(\frac{1}{2}t + \frac{3}{2})$. The optimal control is seen to be

$$u^*(t) = \begin{cases} -1 \text{ (sell)} & 0 \le t < 4.6 \\ 1 \text{ (buy)} & 4.6 < t < 7. \end{cases}$$

The simple optimal policy (12) has some shortcomings, particularly for long planning periods. It is very much dependent on the final price of wheat $p(T)$, and not on the price between t and T. If T is very large, then for small t the function $t \to p(T) - \alpha(T - t)$ will be negative, and (12) would require us to sell wheat. This would mean (for T sufficiently large) that the supply of wheat $x_2(t)$ would eventually become negative, which would gain us cash with the state equation having the form (4). So for long planning periods, we should modify (4) to assign a cost for short-selling (that is, $x_2 < 0$) or else forbid short-selling by adding the state constraint.

$$x_2(t) \geq 0, \qquad t \geq 0.$$

We shall return to this problem in Chapter VIII, when we discuss state-constrained control problems.

3. EXISTENCE OF OPTIMAL CONTROLS

In this section we summarize some of the main existence theorems of optimal control theory. From now on we will assume the set of admissible controls is

$$\Delta = \{\mathbf{u} : \mathbf{u} \text{ is bounded, measurable, and } \mathbf{u}(t) \in \Omega, \text{ for all } t\},$$

for some compact set $\Omega \subset R^m$. In order to have a reasonable existence theory, it is necessary to consider measurable controllers, not just piecewise continuous. This may seem to lead us away from physical reality. On the other hand, once the existence of an optimal control is guaranteed, the necessary conditions, such as the maximum principle, can be applied without qualm, and in many cases they will show that the optimal controls are indeed piecewise continuous or better.

The first result is essentially due to Fillipov. For a detailed proof, we refer to [2].

Theorem 1 Consider the nonlinear control process

$$\dot{\mathbf{x}} = \mathbf{f}(\mathbf{x}, t, \mathbf{u}), \qquad t > 0, \qquad \mathbf{x}(0) \in X_0,$$

where $\mathbf{f}$ is continuously differentiable, and

(a) the initial and target sets X_0, X_1 are nonempty compact sets in R^n;

(b) the control restraint set Ω is compact, and there exists a control transferring X_0 to X_1 in finite time;

(c) the cost for each $\mathbf{u} \in \Delta$ is

$$C(\mathbf{u}) = g(\mathbf{x}(t_1)) + \int_0^{t_1} f_0(\mathbf{x}(t), t, \mathbf{u}(t))\, dt + \max_{t \in [t_0, t_1]} \{\gamma(\mathbf{x}(t))\},$$

where f_0 is continuously differentiable in R^{n+1+m} and g and γ are continuous on R^n.

Assume

(d) there exists a uniform bound

$$|\mathbf{x}(t)| \leq b \quad 0 \leq t \leq t_1,$$

For all responses $\mathbf{x}(t)$ to all controllers $\mathbf{u} \in \Delta$;

(e) the extended velocity set

$$\tilde{V}(\mathbf{x}, t) = \{(f^0(\mathbf{x}, t, \mathbf{u}), \mathbf{f}(\mathbf{x}, t, \mathbf{u}) : \mathbf{u} \in \Omega\}$$

is convex in R^{n+1} for each fixed $(\mathbf{x}, t)$.

Then there exists an optimal controller $\mathbf{u}^*(t)$ on $[0, t_1^*]$, $\mathbf{u}^* \in \Delta$, minimizing $C(\mathbf{u})$.

Corollary 1 For the time optimal problem if (a), (b), (c), (d) hold, and $V(\mathbf{x}, t) = \{f(\mathbf{x}, t, \mathbf{u}) | \mathbf{u} \in \Omega\}$ (the velocity set) is convex in R^n for each fixed $(\mathbf{x}, t)$, then time optimal controls exist.

Applications of the theorem and its corollary are plentiful. Consider our moon-landing problem:

$$\text{minimize} \int_0^{t_1} u(\tau)\, d\tau$$

$$\text{subject to} \quad \dot{h} = v, \quad \dot{v} = -g + m^{-1}u, \quad \dot{m} = -ku.$$

Hence

$$\tilde{V}((h, v, m), t) = \{(u, v, -g + m^{-1}u, -ku) : 0 \leq u \leq 1\},$$

which is convex for each fixed $((h, v, m), t)$. Hence an optimal control for this problem exists, and all our previous computations are justified.

Similarly for Example 3 in Chapter I,

$$\text{minimize} \int_0^{T} (1 - u)x\, dt$$

$$\text{subject to} \quad \dot{x} = kux, \quad 0 \leq u \leq 1.$$

We have

$$\tilde{V}(x, t) = \{((1 - u)x, kux) : 0 \leq u \leq 1\} = \{(x - ux, kux) : 0 \leq u \leq 1\},$$

There are, however, many problems where optimal controls do not exist [1].

Example 1 Consider the time optimal problem of hitting the fixed target (1, 0) with the system

$$\begin{aligned} \dot{x}_1 &= (1 - x_2^2)u^2, \qquad & x_1(0) &= 0, \\ x_2 &= u, & x_2(0) &= 0, \end{aligned} \qquad |u(t)| \leq 1.$$

We show first that $(1,0) \in \overline{\mathscr{A}(1)}$ but $(1,0) \notin \mathscr{A}(1)$.

For each positive integer n, subdivide $[0, 1]$ into $2n$ equal subintervals. Let $I_j = (j/2n, (j+1)/2n), j = 0, 1, \ldots, 2n - 1$. Define

$$u_n(t) = \begin{cases} 1 & \text{if } \; t \in I_j, \; j \text{ odd} \\ -1 & \text{if } \; t \in I_j, \; j \text{ even}. \end{cases}$$

For example (see Fig. 2), let $x(t\,; u_n)$ denote the solution corresponding to u_n. Then

$$x_2(1 : u_n) = \int_0^t u_n(\tau)\, d\tau = 0 \qquad \text{for all } n$$

and

$$|x_2(t : u_n)| \leq \left| \int_0^t u_n(\tau)\, d\tau \right| \to 0 \qquad \text{uniformly in} \quad t \in [0, 1]$$

as $n \to \infty$. Thus

$$\begin{aligned} x_1(1 : u_n) &= \int_0^1 [1 - x_2(\tau : u_n)^2] u_n^2(\tau)\, d\tau \\ &= \int_0^1 (1 - x_2(\tau : u_n(\tau))^2\, d\tau \to 1 \qquad \text{as} \quad n \to \infty. \end{aligned}$$

Hence $x(1 : u_n) \to (1, 0)$; however, $(1, 0) \notin \mathscr{A}(1)$, since this would require a control u such that $x_2(t : u) \equiv 0$, or $u \equiv 0$ and then $x_1(t : u) \equiv 0$.

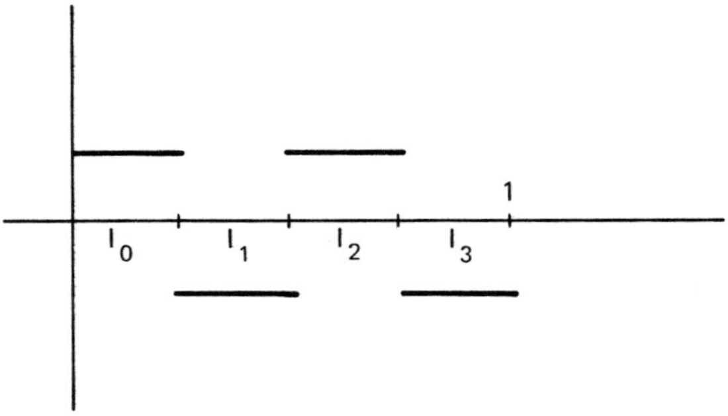

Fig. 2

As an exercise, show that for any $t > 1, (1,0) \in \mathscr{A}(t)$, and hence $\inf\{t : (1,0) \text{ is reached in time } t\} = 1$, but this infimum cannot be attained.

As the final existence theorem we consider the time optimal problem when the state equations are linear in $\mathbf{x}$. In this case we do not need any convexity.

Theorem 2 (Olech and Neustadt [4]) Consider the time optimal control problem with state equation

$$\dot{\mathbf{x}}(t) = A(t)\mathbf{x}(t) + \mathbf{f}(t, \mathbf{u}(t)), \qquad \mathbf{x}(0) = \mathbf{x}_0,$$

where $\mathbf{f}$ is (separately) continuous and the entries of A are integrable over finite intervals. Suppose Δ, X_1 are as before. If the target set X_1 is hit in some time by an admissible control, then it is hit in minimum time by an optimal control.

Problems

1. Use the maximum principle to solve the following problem:

maximize $(8x_1(18) + 4x_2(18))$

$$\text{subject to} \quad \begin{aligned} \dot{x}_1 &= 2x_1 + x_2 + u, & x_1(0) &= x_1^0, \\ \dot{x}_2 &= 4x_1 - 2u, & x_2(0) &= x_2^0, \end{aligned} \qquad 0 \le u \le 1.$$

2. Resolve the control problem in Chapter 3, Problem 5, with the cost

$$\text{maximize } x(100).$$

3. Discuss optimal control of the system

$$\text{minimize } \frac{1}{2}\int_0^{t_1} x_1^2\, dt$$

$$\text{subject to} \quad \begin{aligned} \dot{x}_1 &= x_2 + u, & x_1(0) &= x_1^0, \\ \dot{x}_2 &= -u, & x_2(0) &= x_2^0, \end{aligned} \qquad |u| \le 1,$$

and $\mathbf{x}(t_1) = \mathbf{0}$. (Note that a singular arc $u^* = -x_1^* - x_2^*$, $x_1^* x_2^* + \frac{1}{2}(x_1^*)^2 = \text{const}$, is possible in this example.)

4. Discuss the minimum time control to the origin for the system

$$\dot{x}_1 = x_2, \qquad \dot{x}_2 = x_3, \qquad \dot{x}_3 = u, \qquad \mathbf{x}(0) = \mathbf{x}^0, \qquad |u| \le 1.$$

5. Discuss the possible singular solution of the system

$$\dot{x}_1 = x_2, \qquad |u| \le 1,$$
$$\dot{x}_2 = -x_2 - x_1 u,$$

$\mathbf{x}(0) = \mathbf{x}^0$, with terminal restraint set $X_1 = \{(x_1, x_2) : x_1^2 + x_2^2 = 1\}$, and cost

$$C(u) = \frac{1}{2} \int_0^t (x_1^2(t) + x_2^2(t))\, dt.$$

(Show that the singular solution trajectory is $x_1^* = \pm x_2^*$, and that the optimal control there is

$$u^* = -(x_1^* + x_2^*)/x_1^*.$$

6. Discuss the optimal solution of the fixed time control problem

$$\dot{x}_1 = -x_1 + x_2, \qquad \dot{x}_2 = -2x_2 + u, \qquad |u| \le 1,$$
$$C(u) = \int_0^T (x_1^2 + |u|)\, dt.$$

Include the possibility of singular arcs.

7. Consider the following model for a rocket in plane flight. Let $x_1(t)$ and $x_2(t)$ be the cartesian coordinates of the rocket; $x_3(t) = dx_1/dt$ and $x_4(t) = dx_2/dt$ the velocity components; $x_5(t)$ the mass of the rocket; $u_1(t)$ and $u_2(t)$ the direction cosines of the thrust vector; $u_3(t) = -dx_5/dt$ the mass flow rate; c the effective exhaust gas speed (positive constant); and g the magnitude of gravitational acceleration (positive constant). Define

$$u_1 = \cos\alpha,$$
$$u_2 = \cos\beta.$$

The state equations are then

$$\dot{x}_1 = x_3,$$
$$\dot{x}_2 = x_4,$$
$$\dot{x}_3 = Cu_1u_3/x_5, \qquad \mathbf{x}(0) = \mathbf{x}_0.$$
$$\dot{x}_4 = Cu_2u_3/x_5 - g,$$
$$\dot{x}_5 = -u_3,$$

The control restraint set is defined by

$$u_1^2 + u_2^2 = 1, \qquad 0 \le u_3 \le M.$$

Say as much as you can about the control that transfers the rocket from $\mathbf{x}_0$ to $[\cdot, A, \cdot, \cdot, B]$ $(A, B$ fixed) and minimizes

$$-\int_0^t x_3\,dt.$$

(That is, we wish to transfer a rocket of given initial mass and position to a given altitude while expending a prescribed amount of fuel and attaining *maximum range*.)

8. Show that time optimal control transfering $(0,0)$ to $(1,0)$ does not exist in the example in Section 3.

References

[1] H. Hermes and J. P. LaSalle, "Functional Analysis and Time-Optimal Control." Academic Press, New York, 1969.

[2] E. B. Lee and L. Marcus, "Foundations of Optimal Control Theory." Wiley, New York, 1967.

[3] C. Norström, The continuous wheat trading model reconsidered. An application of mathematical control theory with a state constraint. Working Paper, Graduate School of Industrial Administration, Carnegie-Mellon University, 1978.

[4] C. Olech, Extremal solutions of a control system, *J. Differential Equations* **2**, 74–101 (1966).

Chapter V Numerical Solution of Two-Point Boundary-Value Problems

1. LINEAR TWO-POINT BOUNDARY-VALUE PROBLEMS

We shall consider methods for the numerical solution of the following types of problem:

(i) n first-order equations are to be solved over the interval $[t_0, t_f]$, where t_0 is the initial point, and t_f is the final point;
(ii) r boundary conditions are specified at t_0;
(iii) $(n - r)$ boundary conditions are specified at t_f.

Without loss of generality we shall take the problem in the form

$$\begin{aligned} \dot{y}_i &= g_i(y_1, y_2, \ldots, y_n, t), \qquad & i &= 1, \ldots, n, \\ y_i(t_0) &= c_i, & i &= 1, \ldots, r, \\ y_i(t_f) &= c_i, & i &= r + 1, \ldots, n, \end{aligned} \tag{1}$$

where each g_i is twice differentiable with respect to y_j.

There can be considerable numerical difficulties in solving such problems. It has been shown for

$$\dot{y} = 16 \sinh 16y, \qquad y(0) = y(1) = 0,$$

that if we choose the missing initial condition,

$$y(0) = s,$$

for $s > 10^{-7}$, then the solution goes to infinity at some point in $(0, 1)$. (This point is $\sim\frac{1}{16}\ln(8/s)$.) The true solution is of course $y(x) \equiv 0$, $0 \leq x \leq 1$ [2].

Let us first consider the linear case

$$\dot{\mathbf{y}}(t) = A(t)\mathbf{y}(t) + \mathbf{f}(t), \qquad t_0 \leq t \leq t_f, \tag{2}$$

where

$$y_i(t_0) = c_i, \qquad i = 1, \ldots, r,$$
$$y_i(t_f) = c_i, \qquad i = r + 1, \ldots, n.$$

So $\mathbf{c}$ and $\mathbf{f}(t)$ are given.

The adjoint system is defined as the solution of the homogeneous equation

$$\dot{\mathbf{z}}(t) = -A^{\mathrm{T}}(t)\mathbf{z}(t). \tag{3}$$

As usual we can write the general solution of (2) as

$$\mathbf{y}(t_f) = X(t_f)\mathbf{y}(t_0) + \int_{t_0}^{t_f} X(t_f)X(s)^{-1}\mathbf{f}(s)\,ds \tag{4}$$

where

$$\dot{X} = A(t)X, \qquad X(t_0) = I.$$

Furthermore, the solution of (3) can be written

$$\mathbf{z}(t_f) = X(t_f)^{-\mathrm{T}}\mathbf{z}(t_0) \tag{5}$$

and so (4) gives

$$\mathbf{z}(t_f)^{\mathrm{T}}\mathbf{y}(t_f) = \mathbf{z}(t_f)^{\mathrm{T}}X(t_f)\mathbf{y}(t_0) + \mathbf{z}(t_f)^{\mathrm{T}} \int_{t_0}^{t_f} X(t_f)X(s)^{-1}\mathbf{f}(s)\,ds. \tag{6}$$

Substituting the transpose of (5) into (6) gives

$$\mathbf{z}(t_f)^{\mathrm{T}}\mathbf{y}(t_f) = \mathbf{z}(t_0)^{\mathrm{T}}\mathbf{y}(t_0) + \int_{t_0}^{t_f} \mathbf{z}(t_f)^{\mathrm{T}}X(t_f)X(s)^{-1}\mathbf{f}(s)\,ds.$$

Now, again from (5),

$$\mathbf{z}(s) = X(s)^{-\mathrm{T}}X(t_f)^{\mathrm{T}}\mathbf{z}(t_f),$$

and so

$$\mathbf{z}(t_f)^{\mathrm{T}}\mathbf{y}(t_f) = \mathbf{z}(t_0)^{\mathrm{T}}\mathbf{y}(t_0) + \int_{t_0}^{t_f} \mathbf{z}(s)^{\mathrm{T}}\mathbf{f}(s)\,ds \tag{7}$$

or in component form,

$$\sum_{i=1}^{n} z_i(t_f)y_i(t_f) - \sum_{i=1}^{n} z_i(t_0)y_i(t_0) = \int_{t_0}^{t_f} \sum_{i=1}^{n} z_i(s)f_i(s)\,ds. \tag{8}$$

Equations (7) and (8) are the basic identities in the method of adjoints, the first of the shooting methods we shall investigate.

The method begins by integrating the adjoint equations (3) backward $n - r$ times, with the terminal boundary conditions

$$\mathbf{z}^{(1)}(t_f) = \left[\begin{array}{c} 0 \\ 0 \\ \vdots \\ 0 \\ \hline 1 \\ 0 \\ 0 \\ \vdots \\ 0 \end{array}\right] r\text{th}, \qquad \mathbf{z}^{(2)}(t_f) = \left[\begin{array}{c} 0 \\ 0 \\ \vdots \\ 0 \\ \hline 0 \\ 1 \\ 0 \\ \vdots \\ 0 \end{array}\right], \qquad \mathbf{z}^{(3)}(t_f) = \left[\begin{array}{c} 0 \\ 0 \\ \vdots \\ 0 \\ \hline 0 \\ 0 \\ 1 \\ \vdots \\ 0 \end{array}\right] r\text{th}, \tag{8'}$$

$$\mathbf{z}^{(n-r)}(t_f) = \left[\begin{array}{c} 0 \\ \vdots \\ 0 \\ \hline 0 \\ \vdots \\ 1 \end{array}\right].$$

This gives us $n - r$ functions $\mathbf{z}^{(1)}(t), \mathbf{z}^{(2)}(t), \ldots, \mathbf{z}^{(n-r)}(t)$, $t_0 \le t \le t_f$ (of course in practice we do the integration at a discrete set of points). Then

$$\mathbf{z}^{(m)}(t_f)^{\mathrm{T}}\mathbf{y}(t_f) = \sum_{i=1}^{n} z_i^{(m)}(t_f)y_i(t_f) = y_{r+m}(t_f) = c_{r+m}, \qquad m = 1, \ldots, n - r,$$

and so (8) becomes, with some rearrangement,

$$\sum_{i=r+1}^{n} z_i^{(m)}(t_0)y_i(t_0) = c_{r+m} - \sum_{i=1}^{r} z_i^{(m)}(t_0)y_i(t_0) - \int_{t_0}^{t_f} \sum_{i=1}^{n} z_i^{(m)}(s)f_i(s)\,ds \tag{9}$$

for $m = 1, \ldots, n - r$.

The set of equations (9) is a set of $n - r$ equations in $n - r$ unknowns $\{y_{r+1}(t_0), y_{r+2}(t_0), \ldots, y_n(t_0)\}$, which we can solve and thereby obtain a full set of initial conditions at $t = t_0$.

Note that (9) can be rewritten

$$\begin{bmatrix} z_{r+1}^{(1)}(t_0) & z_{r+2}^{(1)}(t_0) & \cdots & z_n^{(1)}(t_0) \\ z_{r+1}^{(2)}(t_0) & z_{r+2}^{(2)}(t_0) & \cdots & z_n^{(2)}(t_0) \\ \vdots & \vdots & & \vdots \\ z_{r+1}^{(n-r)}(t_0) & z_{r+2}^{(n-r)}(t_0) & \cdots & z_n^{(n-r)}(t_0) \end{bmatrix} \begin{bmatrix} y_{r+1}(t_0) \\ y_{r+2}(t_0) \\ \vdots \\ y_n(t_0) \end{bmatrix}$$

$$= \begin{bmatrix} c_{r+1} - \sum_{i=1}^{r} z_i^{(1)}(t_0)c_i - \int_{t_0}^{t_f} \sum z_i^{(1)}(t) f_i(t)\, dt \\ c_{r+2} - \sum_{i=1}^{r} z_i^{(2)}(t_0)c_i - \int_{t_0}^{t_f} \sum z_i^{(2)}(t) f_i(t)\, dt \\ \vdots \\ c_n - \sum_{i=1}^{r} z_i^{(n-r)}(t_0)c_i - \int_{t_0}^{t_f} \sum z_i^{(n-r)}(t) f_i(t)\, dt \end{bmatrix} \tag{10}$$

The inverse of the z matrix exists since $\{\mathbf{z}^{(1)}(t_f), \ldots, \mathbf{z}^{(n-r)}(t_f)\}$ are linearly independent, and consequently so are

$$\{\mathbf{z}^{(1)}(t), \ldots, \mathbf{z}^{(n-r)}(t)\} \qquad \text{for any} \quad t \in [t_0, t_f].$$

(This is a known fact from linear ordinary differential equations.)

We can code this method as follows:

1. Set $m = 1$.
2. Integrate the adjoint equations backward from t_f to t_0 for the mth set of boundary conditions (8′).
3. Evaluate the mth row of (10).
4. If $m = n - r$, solve Eq. (10) for the missing initial conditions $\{y_{r+1}(t_0), \ldots, y_n(t_0)\}$; go to item 6.
5. *If* $m < n - r$, set $m = m + 1$; return to item 2.
6. Now using the full set of initial conditions $\{y_i(t_0)\}_{i=1}^{n}$, integrate (1) forward to obtain the solution of the boundary-value problem.

Note that if $n - r \geq r$, we save work by reversing the roles of the initial and terminal points.

2. NONLINEAR SHOOTING METHODS

Nonlinear two-point boundary problems can be solved by an iterative process. We start with an initial guess

$$\{y_{r+1}^{(0)}(t_0), y_{r+2}^{(0)}(t_0), \ldots, y_n^{(0)}(t_0)\}.$$

This will allow us to solve the Eqs. (1.1) to find $\mathbf{y}^{(0)}(t)$, $t_0 \le t \le t_f$. Then we iterate according to the following scheme: set

$$\delta\mathbf{y}^{(k)}(t) = \mathbf{y}^{(k+1)}(t) - \mathbf{y}^{(k)}(t) \qquad t_0 \le t \le t_f. \tag{1}$$

Then by the usual Taylor-series expansion, we have as a first-order approximation

$$\delta\dot{y}_i^{(k)}(t) = \sum_{j=1}^{n} \frac{\partial g_i}{\partial y_j}(\mathbf{y}^{(k)}(t))\,\delta y_j^{(k)}(t), \qquad i = 1, \ldots, n,$$

that is,

$$\delta\dot{\mathbf{y}}^{(k)} = J(\mathbf{y}^{(k)})\delta\mathbf{y}^{(k)} \tag{2}$$

(where J is the gradient of $\mathbf{g}$), which is just a set of linear ordinary differential equations with variable coefficients.

Furthermore,

$$\begin{aligned} \delta y_i^{(k)}(t_0) &= 0, & i &= 1, \ldots, r, & k &= 0, 1, 2, \ldots, \\ \delta y_i^{(k)}(t_f) &= c_i - y_i^{(k)}(t_f), & i &= r+1, \ldots, n, & k &= 0, 1, 2, \ldots. \end{aligned} \tag{3}$$

Equations (2) and (3) define a linear two-point boundary-value problem for the correction vector $\delta\mathbf{y}^{(k)}(t)$, $t_0 \le t \le t_f$, which we can solve by the previous method, the method of adjoints; namely, as before, we define the adjoint system to (2)

$$\dot{\mathbf{z}}^{(k)}(t) = -J_k^{\mathrm{T}}(t)\mathbf{z}^{(k)}(t), \tag{4}$$

where $J_k = J(\mathbf{y}^{(k)}(t))$, and solve (4) backward $n - r$ times with end-point conditions

$$r\text{th}\begin{bmatrix} 0 \\ 0 \\ \vdots \\ \underline{0} \\ 1 \\ 0 \\ \\ \vdots \\ 0 \end{bmatrix}, \quad \begin{bmatrix} 0 \\ 0 \\ \vdots \\ \underline{0} \\ 0 \\ 1 \\ \\ \vdots \\ 0 \end{bmatrix}, \quad \begin{bmatrix} 0 \\ 0 \\ \vdots \\ \underline{0} \\ 0 \\ 0 \\ 1 \\ \vdots \\ 0 \end{bmatrix}, \quad \ldots, \quad \begin{bmatrix} 0 \\ \\ \vdots \\ \underline{0} \\ 0 \\ \\ \\ \vdots \\ 1 \end{bmatrix}.$$

Denote the solutions thus obtained at the kth iteration step by

$$\{\mathbf{z}^{(1)}(t), \mathbf{z}^{(2)}(t), \ldots, \mathbf{z}^{(n-r)}(t)\}_{(k)}.$$

Then the fundamental identity tells us that

$$\begin{bmatrix} z_{r+1}^{(1)}(t_0) & z_{r+2}^{(1)}(t_0) & \cdots & z_n^{(1)}(t_0) \\ z_{r+1}^{(2)}(t_0) & z_{r+2}^{(2)}(t_0) & \cdots & z_n^{(2)}(t_0) \\ & \vdots & & \\ z_{r+1}^{(n-r)}(t_0) & z_{r+2}^{(n-r)}(t_0) & \cdots & z_n^{(n-r)}(t_0) \end{bmatrix}_{(k)} \begin{bmatrix} \delta y_{r+1}^{(k)}(t_0) \\ \delta y_{r+2}^{(k)}(t_0) \\ \vdots \\ \delta y_n^{(k)}(t_0) \end{bmatrix} = \begin{bmatrix} \delta y_{r+1}^{(k)}(t_f) \\ \delta y_{r+2}^{(k)}(t_f) \\ \vdots \\ \delta y_n^{(k)}(t_f) \end{bmatrix}. \quad (5)$$

[Note that (2) has no forcing term, simplifying Eqs. (5).]

Solving (5) gives us a complete set of initial conditions $\delta \mathbf{y}^{(k)}(t_0)$, and we get the next choice for the initial conditions for $\mathbf{y}$ by setting

$$\mathbf{y}^{(k+1)}(t_0) = \mathbf{y}^{(k)}(t_0) + \delta \mathbf{y}^{(k)}(t_0).$$

Now we solve for $\mathbf{y}^{(k+1)}(t)$, $t \in [t_0, t_f]$, and return to (2) to calculate $J(\mathbf{y}^{(k+1)})$, $\delta \mathbf{y}^{(k+1)}$, etc.

We terminate whenever

$$\max\{\delta y_i^{(k)}(t_f) : i = r, r+1, \ldots, n\}$$

is sufficiently small or k is too large.

This is called the *shooting method*; we guess the unknown initial values, solve the equations, and then, on the basis of this solution, make corrections to the previous initial values. So in shooting methods our aim is to find the missing initial data.

It can be shown that shooting methods are a special case of the Newton–Raphson method, and so, provided that our initial guess is sufficiently close, J_k is nonsingular, and the interval $[t_0, t_f]$ is not too large, the approximations converge quadratically.

To recapitulate, the method of adjoints for nonlinear ordinary differential equations is carried out as follows:

1. Determine analytically the gradient $(\partial g_i/\partial y_j)$.
2. Initialize the counter on the iterative process. Set $k = 0$.
3. For $k = 0$, guess the missing initial conditions $y_i^{(0)}(t_0)$, $i = r+1, \ldots, n$.
4. Integrate (1) of Section 1 with initial conditions

$$\begin{aligned} y_i^{(k)}(t_0) &= c_i, & i &= 1, 2, \ldots, r, \\ y_i^{(k)}(t_0), & & i &= r+1, \ldots, n, \end{aligned}$$

and store $\mathbf{y}^{(k)}(t)$.

5. Set the counter on the integration of the adjoint equations, $m = 1$.

6. Calculate $z_i^{(m)}(t_0)$, $i = r + 1, \ldots, n$, by integrating the adjoint equations (4) backward from t_f to t_0, with final data $z_i^{(m)}(t_f)$ as in Eq. (8) of Section 1, $i = 1, \ldots, n$.

Note that in this integration the stored profiles $\mathbf{y}^{(k)}(t)$ are used to evaluate the partial derivatives $\partial g_i/\partial y_j$, $i, j = 1, \ldots, n$.

7. For the mth row of (5) form the right-hand side of (5) by subtracting the specified terminal value $y_i(t_f) = c_i$ from the calculated value $y_i^{(k)}(t_f)$, $i = r + 1, \ldots, n$, found in item 4.

8. If $m < n - r$, set $m = m + 1$ and go to 6.

9. Form the set of $n - r$ linear algebraic equations (5) and solve for $\delta y_i^{(k)}(t_0)$, $i = r + 1, \ldots, n$.

10. Form the next set of trial values by

$$y_i^{(k+1)}(t_0) = y_i^{(k)}(t_0) + \delta y_i^{(k)}(t_0), \qquad i = r + 1, \ldots, n.$$

11. Set $k = k + 1$; return to 4.

12. Terminate, whenever $\max\{\delta y_i^{(k)}(t_f) : i = r + 1, \ldots, n\}$ is sufficiently small, or whenever k exceeds a maximum value.

3. NONLINEAR SHOOTING METHODS: IMPLICIT BOUNDARY CONDITIONS

In this section we consider the two-point boundary-value problem of Eqs. (1.1) with implicit boundary conditions that are functions of both the initial and the terminal conditions:

$$q_i(y_1(t_0), y_2(t_0), \ldots, y_n(t_0), y_1(t_f), \ldots y_n(t_f)) = 0, \qquad i = 1, \ldots, n. \quad (1)$$

Let us define the variation in q_i as

$$\delta q_i = q_i^{\text{true}} - q_i^{\text{approx}} = -q_i^{\text{approx}}, \qquad i = 1, 2, \ldots, n, \quad (2)$$

Since $q_i^{\text{true}} = 0$. Up to second order we can approximate δq_i by

$$\delta q_i = \sum_{j=1}^{n} \frac{\partial q_i}{\partial y_j(t_0)} \delta y_j(t_0) + \sum_{j=1}^{n} \frac{\partial q_i}{\partial y_j(t_f)} \delta y_j(t_f), \qquad i = 1, 2, \ldots, n, \quad (3)$$

where

$$\left(\frac{\partial q_i}{\partial y_j(t_0)}\right) \quad \text{and} \quad \left(\frac{\partial q_i}{\partial y_j(t_f)}\right)$$

are the gradients evaluated at $\mathbf{y}(t_0)$ and $\mathbf{y}(t_f)$, respectively. Equations (3) are n equations in $2n$ variables $\{\delta\mathbf{y}(t_0), \delta\mathbf{y}(t_f)\}$. However, from the fundamental identity of adjoints we can relate $\delta\mathbf{y}(t_0)$ and $\delta\mathbf{y}(t_f)$ as in Eq. (2.5)

$$\sum_{i=1}^{n} z_i^{(j)}(t_0)\,\delta y_i(t_0) = \delta y_j(t_f), \qquad j = 1, 2, \ldots, n, \tag{4}$$

where $\{\mathbf{z}^{(1)}, \ldots, \mathbf{z}^{(n)}\}$ are the solutions of the adjoint equations (5.2.4) with terminal conditions

$$\mathbf{z}^{(1)}(t_f) = \begin{bmatrix} 1 \\ 0 \\ \vdots \\ 0 \\ 0 \end{bmatrix}, \qquad \mathbf{z}^{(2)}(t_f) = \begin{bmatrix} 0 \\ 1 \\ 0 \\ \vdots \\ 0 \end{bmatrix}, \qquad \ldots, \qquad \mathbf{z}^{(n)}(t_f) = \begin{bmatrix} 0 \\ 0 \\ \vdots \\ \\ 1 \end{bmatrix} \tag{5}$$

If we substitute (4) into (3), we get

$$\delta q_i = \sum_{j=1}^{n} \frac{\partial q_i}{\partial y_j(t_0)}\,\delta y_j(t_0) + \sum_{j=1}^{n} \frac{\partial q_i}{\partial y_j(t_f)} \left(\sum_{s=1}^{n} z_s^{(j)}(t_0)\,\delta y_s(t_0) \right), \qquad i = 1, 2, \ldots, n. \tag{6}$$

On rearranging (6), we find that

$$\delta q_i = \sum_{p=1}^{n} \left(\frac{\partial q_i}{\partial y_p(t_0)} + \sum_{j=1}^{n} \frac{\partial q_i}{\partial y_j(t_f)}\, z_p^{(j)}(t_0) \right) \delta y_p(t_0), \qquad i = 1, 2, \ldots, n. \tag{7}$$

Equation (7) is a set of n equations in n unknowns $\{\delta\mathbf{y}(t_0)\}$, which can be solved, and from $\delta\mathbf{y}(t_0)$ we get our new guess for the initial data.

To recapitulate:

1. Determine analytically the partial derivatives

$$\left(\frac{\partial g_i}{\partial y_j}\right), \qquad \left(\frac{\partial q_i}{\partial y_j(t_0)}\right), \qquad \left(\frac{\partial q_i}{\partial y_j(t_f)}\right).$$

2. Initialize the counter on the iterative process. Set $k = 0$.
3. For $k = 0$, guess the missing initial conditions

$$y_i^{(0)}(t_0), \qquad i = 1, 2, \ldots, n.$$

4. Integrate (1.1) with initial conditions $y_i^{(k)}(t_0)$ and store the profiles $y_i^{(k)}(t)$, $i = 1, 2, \ldots, n$.

5. Using initial values $y_i^{(k)}(t_0)$ and calculated final values $y_i^{(k)}(t_f)$, evaluate δq_i, $i = 1, 2, \ldots, n$ by (1) and (2).

6. If $\max\{\delta q_1, \delta q_2, \ldots, \delta q_n\}$ is sufficiently small, or if k is greater than the maximum terminate. Otherwise go to 7.

7. For each $y_i(t_f)$ appearing in the implicit boundary conditions (1), integrate the adjoint equations (2.4) backward with terminal data (5). The profiles $y_i^{(k)}(t)$, $i = 1, 2, \ldots, n$, that are stored in 4 are used to evaluate the derivatives $(\partial g_i/\partial y_j)$. Save $\{\mathbf{z}^{(1)}(t_0), \ldots, \mathbf{z}^{(n)}(t_0)\}$.

8. Using the expressions in 1 for $(\partial q_i/\partial y_j)$, evaluate

$$\frac{\partial q_i}{\partial y_j(t_0)}, \qquad \frac{\partial q_i}{\partial y_j(t_f)}, \qquad i, j = 1, 2, \ldots, n,$$

and form the left-hand side of (7).

9. Solve (7) for $\delta y_i(t_0)$ and call the solution $\delta y_i^{(k)}(t_0)$ for the kth iteration step, $i = 1, 2, \ldots, n$.

10. Form the next set of trial conditions

$$y_i^{(k+1)}(t_0) = y_i^{(k)}(t_0) + \delta y_i^{(k)}(t_0), \qquad i = 1, \ldots, n.$$

11. Set $k = k + 1$; return to 4.

4. QUASI-LINEARIZATION

Let us reconsider the system of n nonlinear ordinary differential equations:

$$\dot{y}_i = g_i(y_1, y_2, \ldots, y_n, t), \qquad i = 1, 2, \ldots, n, \tag{1}$$

$$y_i(t_0) = c_i, \qquad i = 1, 2, \ldots, r, \tag{2}$$

$$y_i(t_f) = c_i, \qquad i = r + 1, \ldots, n. \tag{3}$$

Suppose we have the kth nominal solution to Eqs. (1)–(3), $\mathbf{y}^{(k)}(t)$ over $[t_0, t_f]$, in the sense that the *initial* and *terminal conditions* are *satisfied exactly*, but the profiles $\mathbf{y}^{(k)}(t)$ *satisfy* the *differential equation* (1) only *approximately*.

We expand the right-hand side of (1) in a Taylor series up through first-order terms around the nominal solution $\mathbf{y}^{(k)}(t)$; namely, approximate

$$g_i(\mathbf{y}^{(k+1)}) \cong g_i(\mathbf{y}^{(k)}) + J_i(\mathbf{y}^{(k)})(\mathbf{y}^{(k+1)} - \mathbf{y}^{(k)}), \qquad i = 1, \ldots, n, \tag{4}$$

where

$$J_i = \left(\frac{\partial g_i}{\partial y_1}, \frac{\partial g_i}{\partial y_2}, \ldots, \frac{\partial g_i}{\partial y_n}\right)$$

is the ith row of the gradient evaluated at $\mathbf{y}^{(k)}$. Since we want

$$\dot{y}_i^{(k+1)} = g_i(\mathbf{y}^{(k+1)}), \qquad i = 1, \ldots, n, \tag{5}$$

by substituting (4) in (5) we arrive at the ordinary differential equation for $\mathbf{y}^{(k+1)}$

$$\dot{y}_i^{(k+1)}(t) = g_i(y_1^{(k)}(t), \ldots, y_n^{(k)}(t), t) + \sum_{j=1}^{n} \frac{\partial g_i}{\partial y_j}(y_j^{(k+1)}(t) - y_j^{(k)}(t)). \tag{6}$$

On rearranging terms in (6), we have

$$\dot{\mathbf{y}}^{(k+1)}(t) = J(\mathbf{y}^{(k)}(t))\mathbf{y}^{(k+1)}(t) + \mathbf{f}(t), \qquad k = 0, 1, 2, \ldots, \tag{7}$$

where $J(\mathbf{y}^{(k)}(t))$ is an $n \times n$ matrix with elements $\partial g_i / \partial y_j$ evaluated at $\mathbf{y}^{(k)}(t)$, and $\mathbf{f}(t)$ is an $n \times 1$ vector with elements

$$g_i(y_1^{(k)}(t), \ldots, y_n^{(k)}(t), t) - \sum_{j=1}^{n} \frac{\partial g_i}{\partial y_j} y_j^{(k)}(t), \qquad i = 1, 2, \ldots, n.$$

Since we are clamping the boundary conditions at each iteration, we set

$$\begin{aligned} y_i^{(k)}(t_0) &= y_i(t_0) = c_i, \qquad i = 1, 2, \ldots, r, \\ y_i^{(k)}(t_f) &= y_i(t_f) = c_i, \qquad i = r+1, \ldots, n. \end{aligned} \tag{8}$$

Equations (7) and (8) define a linear two-point boundary-value problem that can be solved by the method of adjoints to give the $(k+1)$th approximation to Eqs. (1)–(3).

Theoretically, for a solution to the nonlinear problem, we require

$$\lim_{k \to \infty} y_i^{(k)}(t) = y_i^*(t), \qquad i = 1, 2, \ldots, n, \quad t_0 \le t \le t_f.$$

Numerically it is sufficient for

$$|y_i^{(k+1)}(t) - y_i^{(k)}(t)| < \varepsilon, \qquad i = 1, 2, \ldots, n, \quad t_0 \le t \le t_f.$$

Recapitulating, quasi-linearization consists of the following steps:

1. Linearize the right-hand side of (1) to obtain (7).
2. For $k = 0$, provide nominal profiles $y_1^{(0)}, \ldots, y_n^{(0)}(t)$, $t_0 \le t \le t_f$, that *satisfy* the *boundary conditions*.
3. For the kth iteration, using as nominal profiles $\mathbf{y}^{(k)}(t)$, solve the linear two-point boundary-value problem (7) and (8).
4. Test whether

$$|y_i^{(k+1)}(t) - y_i^{(k)}(t)| < \varepsilon, \qquad i = 1, 2, \ldots, n, \quad t_0 \le t \le t_f.$$

If satisfied, exit; otherwise set $k = k + 1$ and go to 3.

It can be shown that when quasi-linearization *converges* it does so *quadratically*, but again it is a Newton-type method (now in $C_0^1[t_0, t_f]$) and so the *initial guess* is very important.

5. FINITE-DIFFERENCE SCHEMES AND MULTIPLE SHOOTING

Finite-difference schemes have proved successful for numerically unstable problems since the finite-difference equations incorporate both the known initial and final data, as does quasi-linearization. Therefore the solution is constrained to satisfy these boundary conditions. The disadvantages are that the preparation of the problem is arduous and the solution involves solving large numbers of (in general) nonlinear simultaneous equations by say Newton–Raphson or other iterative methods.

To see why this arises, consider a simple linear problem

$$\frac{d^2y}{dt^2} = y + t, \qquad a \leq t \leq b, \tag{1}$$

$$y(a) = c_1, \qquad y(b) = c_2. \tag{2}$$

The interval $[a, b]$ is divided into $N + 1$ intervals of length $h = (b - a)/(N + 1)$, discrete points are given by

$$t_j = t_0 + j\frac{b-a}{N+1}, \qquad j = 0, \ldots, N+1,$$

where $t_0 = a$ and $t_{N+1} = b$. Set $y_j = y(t_j)$. Then if we replace y'' by second central difference

$$y'' = \frac{y_{j+1} - 2y_j + y_{j-1}}{h^2} + O(h^2),$$

and the discrete approximation to (1) is

$$-y_{j-1} + (2 + h^2)y_j - y_{j+1} = -h^2 t_j, \qquad j = 1, 2, \ldots, N. \tag{3}$$

Then (1) and (2) can be approximated by the N equations in N unknowns $\{y_1, \ldots, y_N\}$, that is,

$$\begin{aligned}
(2 + h^2)y_1 - y_2 &= -h^2 t_1 + c_1, \\
-y_1 + (2 + h^2)y_2 - y_3 &= -h^2 t_2, \\
&\vdots \\
-y_{N-1} + (2 + h^2)y_N &= -h^2 t_N + c_2.
\end{aligned}$$

Or in matrix form,

$$\begin{bmatrix} 2+h^2 & -1 & & & \\ -1 & 2+h^2 & -1 & & \\ & \ddots & \ddots & \ddots & \\ & & & & -1 \\ & & & -1 & 2+h^2 \end{bmatrix} \begin{bmatrix} y_1 \\ y_2 \\ \vdots \\ y_{N-1} \\ y_N \end{bmatrix} = \begin{bmatrix} -h^2 t_1 + c_1 \\ -h^2 t_2 \\ \vdots \\ -h^2 t_{N-1} \\ -h^2 t_N + c_2 \end{bmatrix} \tag{4}$$

since $y_0 = y(a) = c_1$, $y_{N+1} = y(b) = c_2$.

Since this problem is linear, we have here a linear set of equations. In general, for nonlinear ordinary differential equations we shall get a set of nonlinear equations that must be solved by iteration.

For example, the central difference approximation to

$$-y'' + y^2 = t, \qquad y(a) = c_1, \quad y(b) = c_2,$$

is

$$-y_{j-1} + 2y_j + h^2 y_j^2 - y_{j+1} = -h^2 t_j, \qquad j = 1, 2, \ldots, N.$$

Finally, for unstable problems a combination of shooting and finite-difference ideas, called *multiple shooting*, has given promising results. Here the interval $[t_0, t_f]$ is divided into equal parts, over each of which a shooting method is applied (see Fig. 1). However, in contrast to finite-difference methods, the number of intervals is usually relatively small.

Note that in solving our general problem by shooting over $[t_0, t_f]$, if we use integration scheme of order p and step size h, the errors in the numerical solution of the differential equations is of the order

$$O(h^p \exp[K_1 |t_0 - t_f|]), \qquad K_1 = \text{const},$$

and so reducing the size of the interval over which we shoot can substantially reduce the errors [1]. For instance, consider the simple

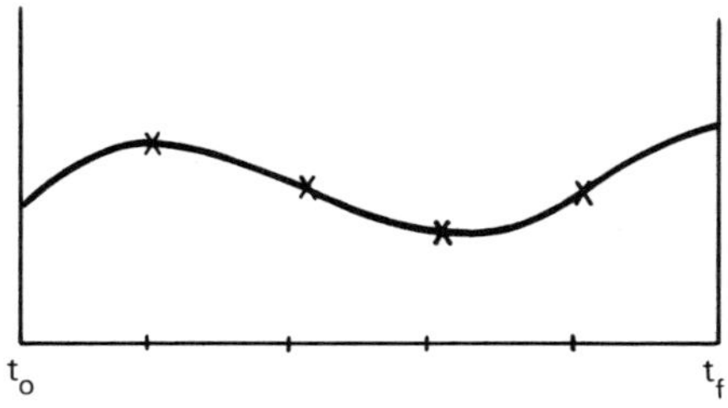

Fig. 1 Multiple Shooting.

two-point boundary-value problem (TPBVP)

$$\begin{bmatrix} \dot{y}_1(t) \\ \dot{y}_2(t) \end{bmatrix} = \begin{bmatrix} f_1(t, \mathbf{y}(t)) \\ f_2(t, \mathbf{y}(t)) \end{bmatrix} \tag{5}$$

over $0 \le t \le 1$, where $y_1(0) = a$, $y_2(1) = b$.

If we break up the interval $[0, 1]$ into two equal subintervals $[0, \frac{1}{2}]$ and $[\frac{1}{2}, 1]$ (see Fig. 2), and apply shooting over each, guessing $y_2(0)$, $y_1(\frac{1}{2})$, $y_2(\frac{1}{2})$, then we can formulate this problem as a four-dimensional TPBVP of a type considered earlier. First, we make a time change to reduce the equation

$$\dot{\mathbf{y}} = f(t, \mathbf{y}), \qquad \tfrac{1}{2} \le t \le 1, \tag{6}$$

to an equation over $[0, \frac{1}{2}]$. That is, if we set $\tau = t - \frac{1}{2}$, $\mathbf{z}(\tau) = \mathbf{y}(\tau + \frac{1}{2})$, $0 \le \tau \le \frac{1}{2}$, then $\mathbf{z}$ is the solution of

$$\dot{\mathbf{z}}(\tau) = \mathbf{f}(\tau + \tfrac{1}{2}, \mathbf{z}(\tau)) := \mathbf{g}(\tau, \mathbf{z}(\tau)), \qquad 0 \le \tau \le \tfrac{1}{2},$$

with

$$z_2(\tfrac{1}{2}) = b. \tag{7}$$

Furthermore, continuity across $t = \frac{1}{2}$ requires that

$$y_1(\tfrac{1}{2}) = z_1(0), \qquad y_2(\tfrac{1}{2}) = z_2(0). \tag{8}$$

Therefore our original problem can be equivalently stated

$$\begin{bmatrix} \dot{y}_1 \\ \dot{y}_2 \\ \dot{z}_1 \\ \dot{z}_2 \end{bmatrix} = \begin{bmatrix} f_1(t, \mathbf{y}) \\ f_2(t, \mathbf{y}) \\ g_1(t, \mathbf{z}) \\ g_2(t, \mathbf{z}) \end{bmatrix}, \qquad 0 \le t \le \tfrac{1}{2}, \tag{9}$$

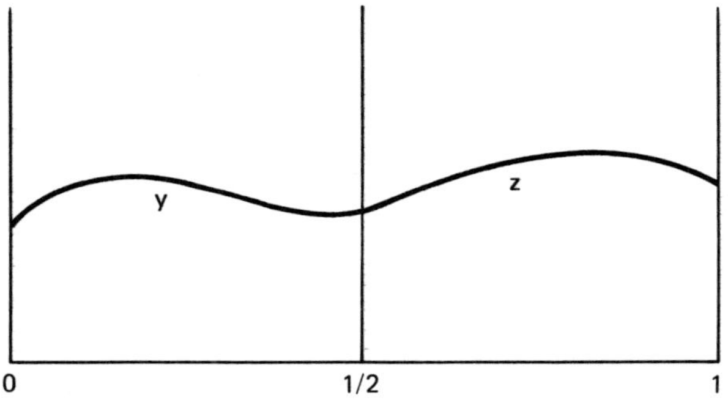

Fig. 2

with boundary conditions (5), (7), (8), or

$$\begin{bmatrix} 1 & 0 & 0 & 0 \\ 0 & 0 & 1 & 0 \\ 0 & 0 & 0 & 1 \\ 0 & 0 & 0 & 0 \end{bmatrix} \begin{bmatrix} y_1(0) \\ y_2(0) \\ z_1(0) \\ z_2(0) \end{bmatrix} + \begin{bmatrix} 0 & 0 & 0 & 0 \\ -1 & 0 & 0 & 0 \\ 0 & -1 & 0 & 0 \\ 0 & 0 & 0 & 1 \end{bmatrix} \begin{bmatrix} y_1(\frac{1}{2}) \\ y_2(\frac{1}{2}) \\ z_1(\frac{1}{2}) \\ z_2(\frac{1}{2}) \end{bmatrix} = \begin{bmatrix} a \\ 0 \\ 0 \\ b \end{bmatrix}, \tag{10}$$

which is of the type considered in Section 3.

Keller [1] has shown that if we partition the interval $[t_0, t_f]$ into N equal subintervals, then we must now solve a nN system of ordinary differential equations by shooting; however, the error can be bounded by

$$O(h^p \exp[K_1|t_f - t_0|]^{1/N}), \qquad K_1 = \text{const.}$$

6. SUMMARY

Method	Advantages	Disadvantages
Shooting	1. Easy to apply. 2. Requires only (N-r) missing initial data points. 3. Converges quadratically when it converges.	1. Stability problems, particularly over long intervals. 2. Must solve nonlinear differential equations.
Quasi-linearization	1. Only need to solve linear differential equations 2. Converges quadratically 3. For numerically sensitive problems quasi-linear equations may be stable.	1. Must select initial profiles. 2. Need to store $y^{(k)}(t)$ and $y^{(k+1)}(t)$.
Finite-Difference Schemes	1. Good for numerically unstable problems.	1. Arduous to code. 2. May need to solve large numbers of nonlinear equations.

Problems

1. Apply the method of adjoints to the equation

$$\ddot{y} = y + t, \qquad y(0) = 0, \quad y(1) = \alpha.$$

2. Apply the method of Section 3 to the nonlinear TPBVP

$$3y\ddot{y} + \dot{y}^2 = 0,$$

with the boundary conditions

$$y(0) = \alpha, \qquad \dot{y}(1) + y^2(1) = \beta$$

3. Write a program to solve the TPBVP

$$\dot{\mathbf{y}}(t) = A(t)\mathbf{y}(t) + \mathbf{f}(t), \qquad 0 \le t \le l$$
$$B\mathbf{y}(0) + C\mathbf{y}(l) = \mathbf{d}, \qquad \mathbf{y}(t) = (y_1(t), \ldots, y_n(t)),$$

for given $A(t)$, $\mathbf{f}(t)$, B, C, and $\mathbf{d}$. As an example, solve

$$\begin{bmatrix} \dot{y}_1 \\ \dot{y}_2 \\ \dot{y}_3 \end{bmatrix} = \begin{bmatrix} 0 & 3 & 1 \\ 1 & 0 & 0 \\ 0 & 2 & 0 \end{bmatrix} \begin{bmatrix} y_1 \\ y_2 \\ y_3 \end{bmatrix} + \begin{bmatrix} e^t \\ 0 \\ t \end{bmatrix}$$

with

$$y_3(0) = -0.75, \qquad y_1(1) = 13.7826, \qquad y_3(1) = 5.64783,$$

over $0 \le t \le 1$, and give the resultant $\mathbf{y}$ values at 0.05 intervals. [2]

References

[1] H. B. Keller, "Numerical Methods for Two-Point Boundary Value Problems. Ginn (Blaisdell), Boston, Massachusetts, 1968.

[2] S. Roberts and J. Shipman, "Two-Point Boundary Value Problems" Elsevier, Amsterdam, 1972.

Chapter VI Dynamic Programming and Differential Games

The fundamental principle of dynamic programming, the so-called *principle* of *optimality*, can be stated quite simply.

If an optimal trajectory is broken into two pieces, then the last piece is itself optimal.

A proof of this principle, sufficient for many applications, can be given quite simply (see Fig. 1). Suppose path (2) is not optimal. Then we can find a "better" trajectory (2′) beginning at (x', t'), which gives a smaller cost than (2). Now, tracing (1) to (x', t') and (2′) from then on must give a smaller overall cost, which contradicts the supposed optimality of (1)–(2). (Of course this proof requires that pieces of admissible trajectory can be concatenated to form an admissible trajectory. This is not necessarily always true, and indeed the principle of optimality may fail to hold.)

1. DISCRETE DYNAMIC PROGRAMMING

Let us consider the network with associated costs shown in Fig. 2. We can regard this as an approximation to a fixed-time problem, where we have partitioned the time interval into three parts, and we have restricted the number of states that are attainable to three after the first time step, three after the second time step, and four after

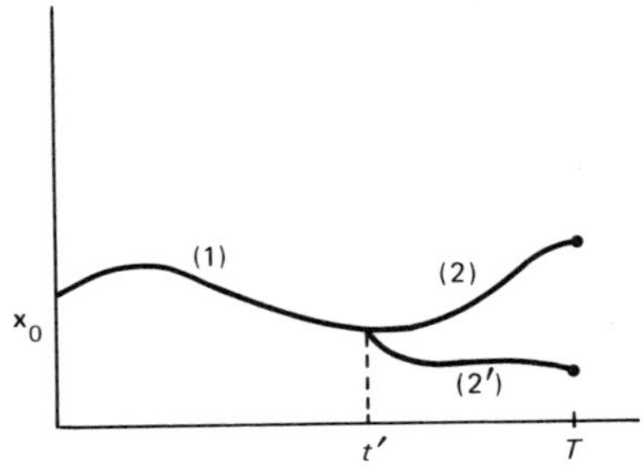

Fig. 1

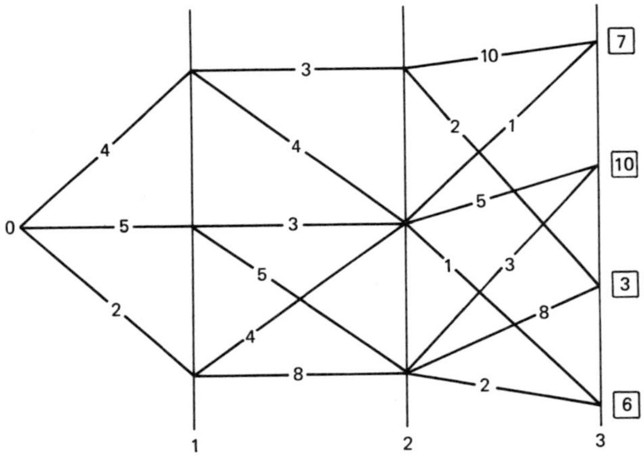

Fig. 2

third. The number on the path refers to the payoff when that path is chosen. Furthermore, not all future states need be attainable from previous states. Finally, the numbers in boxes refer to the extra payoff we get for finishing up at that point. This corresponds to a term $g(x(T))$ in the cost functional. Of course they may be zero.

Let us work out the path that maximizes the total payoff (Fig. 3). Remembering the idea of dynamic programming, we start at the rear (i.e., state 3) and move backward. Now calculate all possible payoffs in going from state 2 to state 3. Remembering that the final section of an optimal trajectory is itself optimal, we can immediately disregard the lower path from P_1 since this gives only 5 units of payoff, whereas the upper path gives 17 units. Another way of looking at this is to

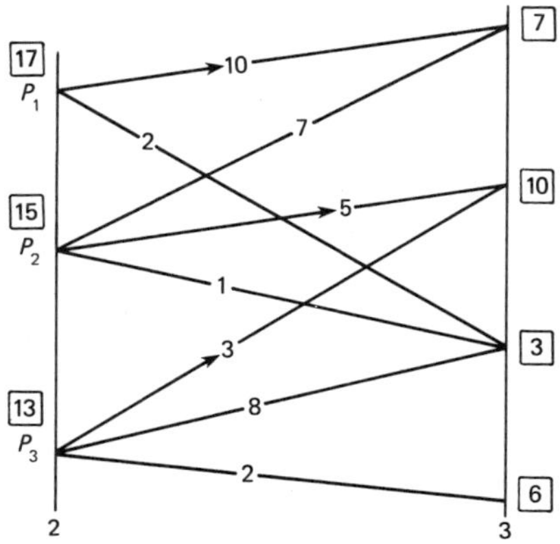

Fig. 3

suppose our optimal trajectory (from 0) lands us at P_1; then clearly over our final stretch we will pick up 17 units and not 5. To indicate that this path is the best from this point, we mark it with an arrow and put 17 next to P_1.

Continuing similarly with P_2 and P_3, we find 15 and 13 as the best payoffs from these points.

Then we apply the same argument in going from step 1 to step 2, and we obtain the result shown in Fig. 4. Note that both paths from the middle point at step 1 give the same cost; hence we arrow both paths.

Now, with one final application (Fig. 5), we have finished. We see that the maximum payoff is 24, and we can follow our arrows backward to get the optimal path: it is the upper path in this case.

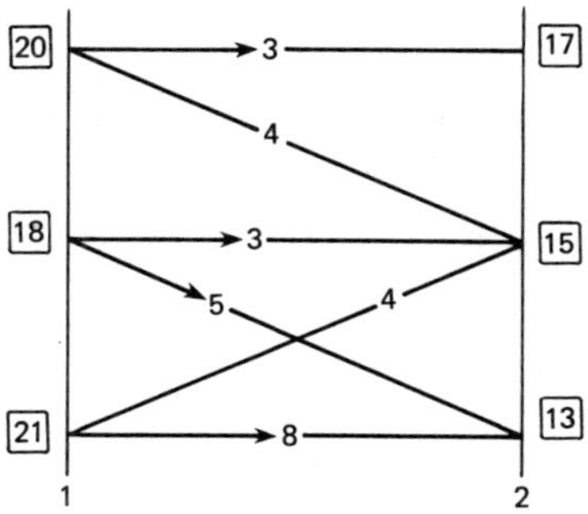

Fig. 4

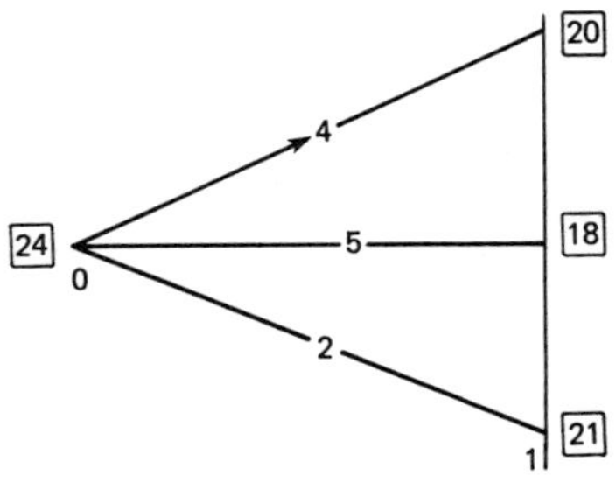

Fig. 5

Unfortunately, if we discretized our time interval into many pieces (n), and allowed many control values at each step (m), then we can see that the dimension of the set of all possible final paths will be enormous (m^n). This is the main disadvantage of dynamic programming: it is often called the *curse of dimensionality*.

2. CONTINUOUS DYNAMIC PROGRAMMING—CONTROL PROBLEMS

We now use dynamic programming to derive necessary conditions for optimality for a control problem of the type

$$\text{minimize} \int_{t_0}^{T} f_0(\mathbf{x}(t), \mathbf{u}(t))\, dt \tag{1}$$

subject to

$$\dot{\mathbf{x}}(t) = \mathbf{f}(\mathbf{x}(t), \mathbf{u}(t)), \qquad \mathbf{x}(t_0) = \mathbf{x}_0, \quad t_0 \leq t \leq T. \tag{2}$$

Let $V(\mathbf{x}_1, t_1)$ be the optimal cost for this control problem with initial condition $\mathbf{x}(t_1) = \mathbf{x}_1$, $t_0 \leq t_1 \leq T$. In the following we shall suppose that V, $\mathbf{f}$, and f_0 are sufficiently differentiable. This is a strong assumption.

Suppose that $\mathbf{u}^*$ is an optimal control over $[t_0, T]$ for (1) and (2), and that $\delta > 0$ is sufficiently small. Then

$$\begin{aligned} V(\mathbf{x}_0, t_0) &= \int_{t_0}^{T} f_0(\mathbf{x}(t : \mathbf{u}^*), \mathbf{u}^*(t))\, dt \\ &= \int_{t_0}^{t_0+\delta} f_0(\mathbf{x}(t : \mathbf{u}^*), \mathbf{u}^*(t))\, dt + \int_{t_0+\delta}^{T} f_0(\mathbf{x}(t : \mathbf{u}^*), \mathbf{u}^*(t))\, dt. \end{aligned} \tag{3}$$

However, by the principle of optimality, the final term in (3) is just $V(\mathbf{x}(t_0 + \delta : \mathbf{u}^*), t_0 + \delta)$, that is,

$$V(\mathbf{x}_0, t_0) = \int_{t_0}^{t_0+\delta} f_0(\mathbf{x}(t : \mathbf{u}^*), \mathbf{u}^*(t))\, dt + V(\mathbf{x}(t_0 + \delta : \mathbf{u}^*), t_0 + \delta). \tag{4}$$

Now, using the regularity of $\mathbf{f}$ and V, we can expand in a Taylor series and see that

$$\mathbf{x}(t_0 + \delta : \mathbf{u}^*) = \mathbf{x}_0 + \delta \mathbf{f}(\mathbf{x}_0, \mathbf{u}^*(t_0)) + o(\delta), \tag{5}$$

$$\begin{aligned} V(\mathbf{x}(t_0 + \delta : \mathbf{u}^*), t_0 + \delta) = \delta\{\nabla_x V(\mathbf{x}_0, t_0) \cdot \mathbf{f}(\mathbf{x}_0, \mathbf{u}^*(t_0)) \\ + V_t(\mathbf{x}_0, t_0)\} + V(\mathbf{x}_0, t_0) + o(\delta), \end{aligned} \tag{6}$$

$$\int_{t_0}^{t_0+\delta} f_0(\mathbf{x}(t : \mathbf{u}^*), \mathbf{u}^*(t))\, dt = \delta f_0(\mathbf{x}_0, \mathbf{u}^*(t_0)) + o(\delta), \tag{7}$$

where

$$\lim_{\delta \to 0}(o(\delta)/\delta) = 0.$$

If Eqs. (5)–(7) are substituted in (4) and $\delta \to 0$, we arrive at the following partial differential equation for V:

$$V_t(\mathbf{x}_0, t_0) = -\{\nabla_x V(\mathbf{x}_0, t_0) \cdot \mathbf{f}(\mathbf{x}_0, \mathbf{u}^*(t_0)) + f_0(\mathbf{x}_0, \mathbf{u}^*(t_0))\}. \tag{8}$$

If we return to Eq. (3), we can derive a little more information about the optimal control. Thus

$$V(\mathbf{x}_0, t_0) = \min_{\mathbf{u} \in \Delta} \left\{ \int_{t_0}^{t_0+\delta} f_0(\mathbf{x}(t : \mathbf{u}), \mathbf{u}(t))\, dt + V(\mathbf{x}(t_0 + \delta : \mathbf{u}), t_0 + \delta) \right\}.$$

Now expanding as before in a Taylor series gives

$$V_t(\mathbf{x}_0, t_0) = -\min_{\mathbf{u}(t_0) \in \Omega} \{\nabla_x V(\mathbf{x}_0, t_0) \cdot \mathbf{f}(\mathbf{x}_0, \mathbf{u}(t_0)) + f_0(\mathbf{x}_0, \mathbf{u}(t_0))\}. \tag{9}$$

Since $\mathbf{u}(t_0)$ can take any value in Ω, and since $\mathbf{x}_0$ and t_0 are arbitrary, (9) implies that

$$V_t(\mathbf{x}, t) = -\min_{\mathbf{w} \in \Omega} \{\nabla_x V(\mathbf{x}, t) \cdot \mathbf{f}(\mathbf{x}, \mathbf{w}) + f_0(\mathbf{x}, \mathbf{w})\}, \tag{10}$$

for all $\mathbf{x}$, $t > 0$, and the optimal control $\mathbf{u}^*$ is just the value of $\mathbf{w} \in \Omega$ that achieves the minimum in (10); $\mathbf{u}^* = \mathbf{u}^*(\mathbf{x}, t)$. Equation (10) is called the Hamilton–Jacobi–Bellman (HJB) equation, and it immediately leads to an optimal control in feedback form.

The partial differential equation (10) has the natural terminal condition

$$V(\mathbf{x}, T) = 0 \qquad \mathbf{x} \in R^n. \tag{11}$$

When the control problem has a cost of the form

$$\text{minimize} \left\{ g(\mathbf{x}(T)) + \int_{t_0}^{T} f_0(\mathbf{x}, \mathbf{u})\, dt \right\},$$

Eq. (11) is replaced by

$$V(\mathbf{x}, T) = g(\mathbf{x}), \qquad \mathbf{x} \in R^n.$$

In general it can be shown that the following theorem holds [3].

Theorem 1 Suppose that $\mathbf{f}$ and f_0 are continuously differentiable and the solution V of (10) is twice continuously differentiable. If $\mathbf{u}^*$ achieves the minimum in (10) and $\mathbf{u}^*$ is piecewise continuous, then $\mathbf{u}^*$ is an optimal control. Conversely, if there exists an optimal control for all $(\mathbf{x}_0, t_0)$, $0 \le t_0 < T$, and V (the optimal cost) is twice continuously differentiable, then V satisfies (10).

Note that if we set

$$\boldsymbol{\psi}(t) = -\nabla_x V(\mathbf{x}^*(t), t), \tag{12}$$

evaluated along an optimal trajectory $\mathbf{x}^*$, then

$$\begin{aligned}
\dot{\psi}_i(t) &= -\frac{d}{dt}\left(\frac{\partial V}{\partial x_i}(\mathbf{x}^*(t), t)\right) \\
&= -\left\{\sum_{j=1}^{n} \frac{\partial^2 V}{\partial x_i \partial x_j}(\mathbf{x}^*(t), t)\frac{dx_j^*(t)}{dt} + \frac{\partial^2 V}{\partial x_i \partial t}(\mathbf{x}^*(t), t)\right\} \\
&= -\sum_{j=1}^{n} \frac{\partial^2 V}{\partial x_i \partial x_j}(\mathbf{x}^*(t), t) f_j(\mathbf{x}^*(t), \mathbf{u}^*(t)) \\
&\quad + \frac{\partial}{\partial x_i}\left\{\sum_{j=1}^{n} \frac{\partial V}{\partial x_j}(\mathbf{x}^*(t), t) f_j(\mathbf{x}^*(t), \mathbf{u}^*(t)) + f_0(\mathbf{x}^*(t), \mathbf{u}^*(t))\right\} \\
&= \sum_{j=1}^{n} \frac{\partial V}{\partial x_j}(\mathbf{x}^*(t), t)\frac{\partial f_j}{\partial x_i}(\mathbf{x}^*(t), \mathbf{u}^*(t)) + \frac{\partial f_0}{\partial x_i}(\mathbf{x}^*(t), \mathbf{u}^*(t)),
\end{aligned}$$

using (10) and the state equation. In other words,

$$\dot{\psi}_i = -\sum_{j=1}^{n} \psi_j \frac{\partial f_j}{\partial x_i}(\mathbf{x}^*, \mathbf{u}^*) + \frac{\partial f_0}{\partial x_i}(\mathbf{x}^*, \mathbf{u}^*), \qquad i = 1, 2, \ldots, n, \tag{13}$$

with $\boldsymbol{\psi}(T) = 0$. Of course Eqs. (13) are the adjoint equations. The other statements of the maximum principle follow analogously; thus under the rather strong regularity assumptions on V, we have derived the maximum principle. The relation (12) also gives us a natural interpretation of the adjoint variables—they are the negative of the rate of change of the cost with respect to x; that is, $-\psi_i$ is the *marginal cost* of x_i.

As an example, let us reconsider the control problem

$$\text{minimize} \int_0^T (ax^2 + u^2)\,dt$$

subject to

$$\dot{x} = \alpha x + \beta u, \qquad x(0) = x_0.$$

The HJB equation becomes

$$\frac{\partial V}{\partial t} = -\min_w \left\{ ax^2 + w^2 + \frac{\partial V}{\partial x}(\alpha x + \beta w) \right\}, \qquad V(x, T) = 0. \quad (14)$$

Since the minimum of the function $h(w) = w^2 + \beta(\partial V/\partial x)w$ occurs at $u^* = -(\beta/2)\,\partial V/\partial x$, (14) becomes

$$-\frac{\partial V}{\partial t} = ax^2 - \frac{\beta^2}{4}\left(\frac{\partial V}{\partial x}\right)^2 + \alpha x\left(\frac{\partial V}{\partial x}\right), \qquad (15)$$

which, when solved for V, gives the optimal cost and optimal control.

We try to compute a solution of the form

$$V(x, t) = c(t)x^2,$$

and we find

$$\dot{c} = \beta^2 c^2 - 2\alpha c - a, \qquad c(T) = 0,$$

which can be solved as before.

As one of the most useful applications of dynamic programming, consider the general quadratic cost linear control problem

$$\text{minimize}\left\{\frac{1}{2}\,\mathbf{x}(T)\cdot S\mathbf{x}(T) + \frac{1}{2}\int_0^T \{\mathbf{x}(t)\cdot Q\mathbf{x}(t) + \mathbf{u}(t)\cdot R\mathbf{u}(t)\}\,dt\right\} \quad (16)$$

subject to

$$\dot{\mathbf{x}}(t) = A\mathbf{x}(t) + B\mathbf{u}(t), \qquad \mathbf{x}(0) = \mathbf{x}_0, \quad 0 \le t \le T. \qquad (17)$$

Here T is a given fixed final time, Q and S are nonnegative definite $n \times n$ matrices, and R is a positive definite $m \times m$ matrix. As controls we allow any square integrable function $\mathbf{u}(t) \in R^m$.

Applying the HJB equation, we find that

$$V_t(\mathbf{x}, t) = -\min_{\mathbf{w}\in R^m}\{\nabla_x V(\mathbf{x}, t)\cdot(A\mathbf{x} + B\mathbf{w}) + \tfrac{1}{2}\mathbf{x}\cdot Q\mathbf{x} + \tfrac{1}{2}\mathbf{w}\cdot R\mathbf{w}\}. \quad (18)$$

A little computation shows that the minimum occurs at

$$\mathbf{u}^*(\mathbf{x}, t) = -R^{-1}B^{\mathrm{T}}\nabla_x V(\mathbf{x}, t), \qquad 0 \le t \le T, \qquad (19)$$

and hence the HJB equation becomes

$$V_t = -\tfrac{1}{2}\mathbf{x}\cdot Q\mathbf{x} - \tfrac{1}{2}\nabla_x V\cdot BR^{-\mathrm{T}}B^{\mathrm{T}}\nabla_x V \\ + \nabla_x V\cdot BR^{-1}B^{\mathrm{T}}\nabla_x V - \nabla_x V\cdot A\mathbf{x}, \qquad (20)$$

where $V(\mathbf{x}, T) = \frac{1}{2}\mathbf{x}\cdot S\mathbf{x}$. By analogy with the previous example, we may be led to seek a solution of (20) in the form

$$V(\mathbf{x}, t) = \tfrac{1}{2}\mathbf{x}\cdot K(t)\mathbf{x} \qquad (21)$$

and, in fact, we find such a solution indeed exists if $K(t)$ satisfies the *matrix Riccati equation*,

$$\dot{K}(t) = -K(t)A - A^{\mathrm{T}}K(t) + K(t)BR^{-1}B^{\mathrm{T}}K(t) - Q(t), \quad 0 \le t \le T, \qquad (22)$$

$K(T) = S$. The existence and uniqueness of such a K is guaranteed by the existence and uniqueness of optimal controls for problem (16)–(17) and by Theorem 1. Of course, in this example, V has the required differentiability properties.

Once the function K has been computed and stored (in practice, of course, this would be done at some discrete set of points in $[0, T]$) the *optimal control* law

$$\mathbf{u}^*(\mathbf{x}, t) = -R^{-1}B^{\mathrm{T}}K(t)\mathbf{x}, \qquad (23)$$

is completely determined in feedback form. Finally, the optimum trajectory is the solution of

$$\dot{\mathbf{x}}^*(t) = [A - BR^{-1}B^{\mathrm{T}}K(t)]\mathbf{x}^*(t), \qquad \mathbf{x}^*(0) = \mathbf{x}_0, \qquad (24)$$

and the optimal cost from $\mathbf{x}_0$ is

$$C(\mathbf{u}^*) = \tfrac{1}{2}\mathbf{x}_0\cdot K(0)\mathbf{x}_0. \qquad (25)$$

Example 1 Suppose that we are given an ideal response $\xi(t)$ for the system (17) and that we wish to minimize the deviation of the actual response $\mathbf{x}(t)$ from the ideal $\xi(t)$ over some period $[0, T]$. Furthermore, we wish to use as small a control energy as possible. Setting

$$\mathbf{e}(t) = \mathbf{x}(t) - \xi(t), \qquad 0 \le t \le T, \qquad (26)$$

our cost could be

$$C(\mathbf{u}) = \mathbf{e}(T)\cdot S\mathbf{e}(T) + \frac{1}{2}\int_0^T (\mathbf{e}(t)\cdot Q\mathbf{e}(t) + \mathbf{u}(t)\cdot R\mathbf{u}(t))\,dt; \qquad (27)$$

$\mathbf{e}(t)$ then represents our "*error*" at time t; the matrices S, Q, and R are chosen to weight the respective terms of the cost functional according to our priorities.

Since $\dot{\mathbf{x}} = A\mathbf{x} + B\mathbf{u}$, substituting in (1) we see that

$$\dot{\mathbf{e}} = A\mathbf{e} + B\mathbf{u} + \mathbf{w}(t), \tag{28}$$

where $\mathbf{w}(t) = A\xi(t) - \dot{\xi}(t)$ is a fixed known function, and

$$\mathbf{e}(0) = \mathbf{x}_0 - \xi(0). \tag{29}$$

The control problem defined by (27)–(29) is analogous to the type considered previously [Eq. (28) has a further term in the right-hand side, but this only slightly changes the argument]. As before, we can derive the following theorem.

Theorem 2 If $R > 0$ and $S, Q \geq 0$, an optimal control exists, is unique, and is given by

$$\mathbf{u}^*(t) = R^{-1}B^{\mathrm{T}}(\mathbf{g}(t) - K(t)\mathbf{x}^*(t)), \tag{30}$$

where K is the solution of the Riccati differential equation

$$\dot{K} = -KA - A^{\mathrm{T}}K + KBR^{-1}B^{\mathrm{T}}K - Q, \qquad K(T) = S. \tag{31}$$

The vector $\mathbf{g}(t)$ is the solution of the linear equation

$$\dot{\mathbf{g}} = -(A - BR^{-1}B^{\mathrm{T}}K)^{\mathrm{T}}\mathbf{g} - Q\xi(t), \qquad \mathbf{g}(T) = S\xi(T). \tag{32}$$

The optimal trajectory satisfies

$$\dot{\mathbf{x}}^*(t) = [A - BR^{-1}B^{\mathrm{T}}K]\mathbf{x}^*(t) + (BR^{-1}B^{\mathrm{T}})\mathbf{g}(t), \qquad \mathbf{x}(0) = \mathbf{x}_0. \tag{33}$$

The minimum value of the cost is

$$C(\mathbf{u}^*) = \tfrac{1}{2}\mathbf{x}^*(T) \cdot S\mathbf{x}^*(T) - \xi(T) \cdot S^{\mathrm{T}}\mathbf{x}^*(T) + \xi(T) \cdot K(T)\xi(T). \tag{34}$$

Example 2 An Infinite Horizon Control Problem (Lee and Marcus [3]) When the basic interval $[0, T]$ becomes infinite, that is, $T = +\infty$, the theory given above leads to the linear regulator problem: find the control that minimizes the total error over all time. To simplify the analysis, we consider the linear autonomous system

$$\dot{\mathbf{x}} = A\mathbf{x} + B\mathbf{u}, \qquad \mathbf{x}(0) = \mathbf{x}_0, \tag{35}$$

for A and B constant matrices, and cost functional

$$C(\mathbf{u}) = \frac{1}{2}\int_0^\infty \{\mathbf{x}(t) \cdot Q\mathbf{x}(t) + \mathbf{u}(t) \cdot R\mathbf{u}(t)\}\, dt, \tag{36}$$

where $Q, R > 0$ are constant symmetric matrices.

The first problem that immediately arises is finding when (36) will be finite. Clearly we want the solutions of (35) (or at least the optimum trajectory) to decay to zero as $t \to \infty$. The set of admissible controls will be the space of all m-vector functions that are square integrable over $(0, \infty)$. It turns out that the required assumption on A and B is

$$\operatorname{rank}\{B, AB, A^2B, \ldots, A^{n-1}B\} = n. \tag{37}$$

(This condition is called controllability, and we shall consider it in some detail in Chapter VII.)

Theorem 3 Consider the autonomous linear control problem (35) for which (37) holds, and the cost functional (36).

There exists a unique symmetric positive definite matrix E such that

$$-A^{\mathrm{T}}E - EA + EBR^{-1}B^{\mathrm{T}}E = Q. \tag{38}$$

For each initial point $\mathbf{x}_0$ in R^n, there exists a unique optimal control $\mathbf{u}^*$ that can be synthesized in feedback form as

$$\mathbf{u}^*(t) = -R^{-1}B^{\mathrm{T}}E\mathbf{x}^*(t). \tag{39}$$

The optimal response $\mathbf{x}^*$ satisfies

$$\dot{\mathbf{x}}^* = (A - BR^{-1}B^{\mathrm{T}}E)\mathbf{x}^*, \qquad \mathbf{x}^*(0) = \mathbf{x}_0, \tag{40}$$

and the minimal cost is

$$C(\mathbf{u}^*) = \tfrac{1}{2}\mathbf{x}_0 \cdot E\mathbf{x}_0. \tag{41}$$

Note that once the existence of E is assumed (basically this involves showing that $\lim_{t\to\infty} K(t) \cdot E$ exists, using the controllability assumption), then by a well-known lemma of Liapunov, the matrix $A - BR^{-1}B^{\mathrm{T}}E$ is a stability matrix (all its eigenvalues have negative real parts), and so solutions of (40) decay to zero as $t \to \infty$, as required [3].

3. CONTINUOUS DYNAMIC PROGRAMMING—DIFFERENTIAL GAMES

In many situations we should like to consider two individually acting players in our control model. Associated with the process would again be a cost functional; however, the players will now be antagonistic, in the sense that one player would try to maximize this cost and the other to minimize it. If the dynamics of the process are again given by a differential equation, we could model it as

$$\dot{\mathbf{x}} = \mathbf{f}(\mathbf{x}, \boldsymbol{\phi}, \boldsymbol{\psi}), \qquad \mathbf{x}(0) = \mathbf{x}_0, \tag{1}$$

where $\boldsymbol{\phi}$ is the first player's control variable and $\boldsymbol{\psi}$ is the second player's control variable. We assume each of these controls takes its values in some restraint set; typically

$$-1 \leq \phi_i \leq 1, \qquad -1 \leq \psi_i \leq 1, \qquad i = 1, \ldots, m \tag{2}$$

or $\boldsymbol{\phi}$ and $\boldsymbol{\psi}$ could be unrestrained [2].

To determine when the game ends, we suppose some terminal set F in the state space R^n is given, and we terminate the game whenever F is reached. To simplify the analysis, we always assume that F can be reached from any point in the state space by admissible controls. Given any two controls $\boldsymbol{\phi}$ and $\boldsymbol{\psi}$, we associate a cost of the form

$$C(\boldsymbol{\phi}, \boldsymbol{\psi}) = \int_0^{t_f} f_0(\mathbf{x}(t), \boldsymbol{\phi}(t), \boldsymbol{\psi}(t))\, dt + g(\mathbf{x}(t_f)), \tag{3}$$

where the integration extends over the trajectory and ends ($t = t_f$) when we hit F. The function g is a terminal cost, which needs to be defined only on F. (We shall show later that this formulation actually includes the fixed-time problem considered earlier.)

The novel aspect is that now $\boldsymbol{\phi}$ will try to minimize (44) and $\boldsymbol{\psi}$ will try to maximize it.

The *value* of the game starting from $\mathbf{x}_0$ would then be naturally defined as

$$V(\mathbf{x}_0) = \min_{\boldsymbol{\phi}} \max_{\boldsymbol{\psi}} C(\boldsymbol{\phi}, \boldsymbol{\psi}), \tag{4}$$

where the minimum and maximum are over all admissible controls $\boldsymbol{\phi}$ and $\boldsymbol{\psi}$. Several mathematical problems involved in such a definition as (4) are immediately obvious. Is the $\min_{\boldsymbol{\phi}} \max_{\boldsymbol{\psi}}$ well defined, and does it equal $\max_{\boldsymbol{\phi}} \min_{\boldsymbol{\psi}}$? A full investigation of these problems involves deep concepts (see [1]), and we shall from here on assume the value exists, and min max = max min whenever required. In the simple problems we shall solve here, this will not be a difficulty.

We can derive necessary conditions for optimality by using dynamic programming in essentially the same way as before. Assume we have a trajectory starting at $\mathbf{x}_0$ at time t_0, of the following type (see Fig. 6). We break the trajectory at time $t_0 + \delta$ for small $\delta > 0$, and suppose that we use arbitrary admissible controls $\boldsymbol{\phi}, \boldsymbol{\psi}$ over $[t_0, t_0 + \delta)$, and optimal controls over the final arc, $t_0 + \delta \to F$. Then the cost calculated

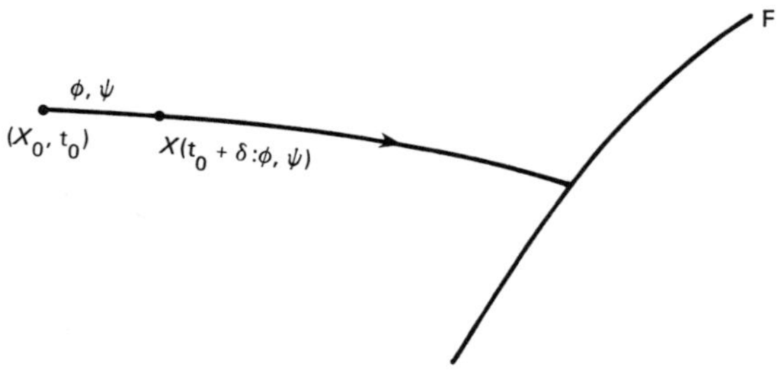

Fig. 6

along this trajectory is just

$$\int_{t_0}^{t_0+\delta} f_0(\mathbf{x}(\tau), \boldsymbol{\phi}(\tau), \boldsymbol{\psi}(\tau))\, dt + V(\mathbf{x}(t_0 + \delta : \boldsymbol{\phi}, \boldsymbol{\psi})),$$

where $\mathbf{x}(t : \boldsymbol{\phi}, \boldsymbol{\psi})$ is the solution of (1) at time t, having used controls $\boldsymbol{\phi}$ and $\boldsymbol{\psi}$. Again expanding $\mathbf{x}(t_0 + \delta : \boldsymbol{\phi}, \boldsymbol{\psi})$ and $V(\mathbf{x}(t_0 + \delta : \boldsymbol{\phi}, \boldsymbol{\psi}))$ by Taylor's theorem gives

$$V(\mathbf{x}(t_0 + \delta : \boldsymbol{\phi}, \boldsymbol{\psi})) = V(\mathbf{x}_0) + \delta \nabla V(\mathbf{x}_0) \cdot \mathbf{f}(\mathbf{x}_0, \boldsymbol{\phi}(t_0), \boldsymbol{\psi}(t_0)) + O(\delta). \quad (5)$$

In other words, the cost of *this* trajectory from $\mathbf{x}_0$ is

$$V(\mathbf{x}_0) + \delta\left(\nabla V \cdot \mathbf{f}(\mathbf{x}_0, \boldsymbol{\phi}(t_0), \boldsymbol{\psi}(t_0)) + \frac{1}{\delta} \int_{t_0}^{t_0+\delta} f_0(\mathbf{x}, \boldsymbol{\phi}, \boldsymbol{\psi})\, d\tau + \frac{O(\delta)}{\delta} \right). \quad (6)$$

Since the value of the game starting at $\mathbf{x}_0$ is the $\min_{\boldsymbol{\phi}} \max_{\boldsymbol{\psi}}$ of the cost (6) (by the principle of optimality), we have

$$V(\mathbf{x}_0) = \min_{\boldsymbol{\phi}} \max_{\boldsymbol{\psi}} [\text{Eq (6)}],$$

so canceling $V(\mathbf{x}_0)$ and letting $\delta \to 0$ gives

$$\min_{\boldsymbol{\phi}} \max_{\boldsymbol{\psi}} [\nabla \mathbf{V} \cdot \mathbf{f}(\mathbf{x}_0, \boldsymbol{\phi}(t_0), \boldsymbol{\psi}(t_0)) + f_0(\mathbf{x}_0, \boldsymbol{\phi}(t_0), \boldsymbol{\psi}(t_0))] = 0. \quad (7)$$

Since (7) holds for any $\mathbf{x}_0$ in the state space and for $\boldsymbol{\phi}(t_0)$ and $\boldsymbol{\psi}(t_0)$ in their respective restraint sets, we have

$$\min \max [\nabla V(\mathbf{x}) \cdot \mathbf{f}(\mathbf{x}, \boldsymbol{\phi}, \boldsymbol{\psi}) + f_0(\mathbf{x}, \boldsymbol{\phi}, \boldsymbol{\psi})] = 0 \quad (8)$$

The minimum and maximum extend over the restraint sets of the controls $\boldsymbol{\phi}$ and $\boldsymbol{\psi}$, respectively. Of course we again have the natural boundary condition for (8),

$$V(\mathbf{x}) = g(\mathbf{x}) \qquad \mathbf{x} \in F. \quad (9)$$

In the same way as before, we can show that the minimum and maximum are obtained by optimal controls.

The fixed-time free-end-point problem considered before can easily be set in this formulation, namely, if T is the terminal time, we define a new *state variable* x_{n+1} by

$$\dot{x}_{n+1}(t) = 1, \qquad x_{n+1}(0) = 0$$

[i.e., $x_{n+1}(t) = t$], and then let F be the set in R^{n+1},

$$F = \{\mathbf{x} \in R^{n+1};\ \mathbf{x} = (x_1, x_2, \ldots, x_n, T),\ \text{all } x_i \in R\}. \tag{10}$$

The state equations now become

$$\begin{bmatrix} \dot{x}_1 \\ \dot{x}_2 \\ \vdots \\ \dot{x}_n \\ \dot{x}_{n+1} \end{bmatrix} = \begin{bmatrix} f_1(\mathbf{x}, \boldsymbol{\phi}, \boldsymbol{\psi}) \\ \vdots \\ f_n(\mathbf{x}, \boldsymbol{\phi}, \boldsymbol{\psi}) \\ 1 \end{bmatrix}, \qquad \mathbf{x}(0) = \mathbf{x}_0. \tag{11}$$

If we substitute (10) into (8), we see

$$V_t = -\min_{\boldsymbol{\phi}} \max_{\boldsymbol{\psi}} \left(\sum_{j=1}^{n} \frac{\partial V}{\partial x_j} f_j(\mathbf{x}, \boldsymbol{\phi}, \boldsymbol{\psi}) + f_0(\mathbf{x}, \boldsymbol{\phi}, \boldsymbol{\psi}) \right), \tag{12}$$

$$V(\mathbf{x}, T) = g(\mathbf{x}), \qquad \mathbf{x} \in R^n$$

the analog of HJB equation derived earlier. We can show that if V is sufficiently differentiable, then this theorem is also sufficient; see, for example, [2]. Similarly, if we let the Hamiltonian

$$H = -[\nabla V \cdot \mathbf{f} + f_0], \tag{13}$$

then, when we substitute in the optimal trajectory for $\mathbf{x}$ and so express V as a function of t, the adjoint equations

$$\frac{d}{dt}(V_{x_i}) = \frac{\partial H}{\partial x_i}(\mathbf{x}^*(t), \boldsymbol{\phi}^*(t), \boldsymbol{\psi}^*(t)), \qquad i = 1, 2, \ldots, n, \tag{14}$$

hold.

Example 1 A War of Attrition and Attack (Isaacs [2]) In the war, we have two opposing sides considered as players I and II in our game-theory model. At time t, player I has supply of vital weapon $x_1(t)$, and player II has supply of his vital weapon $x_2(t)$. Each player can allocate his stock to one of two uses:

(1) *Attrition.* The long-range objective of depleting his enemy's rate of weapon supply.

(2) *Attack.* Entering his weapon supply into a major conflict in the hope of winning the war outright. (This is a short-range policy, and we shall assume that it is the accumulation of these entries that count; each player seeks more than his opponent, and the excess will be the payoff.)

The basic decision is then a choice between a long-range policy of attrition (e.g., guerilla warfare) and a short-range one of direct attack when we think the enemy is sufficiently weakened. Suppose, at time t, player I splits his forces into the *attacking component* $(1 - \phi(t))x_1(t)$ and the *attrition component* $\phi(t)x_1(t)$. His control variable ϕ is then restrained by $0 \leq \phi(t) \leq 1$, $t > 0$. Similarly, player II's control variable will be ψ, $0 \leq \psi(t) \leq 1$, $t > 0$.

Suppose player II, if left unhindered, can manufacture weapons at a rate m_2. He also loses them at a rate proportional to the number ϕx_1 that his enemy is devoting to that purpose. We shall assume therefore

$$\dot{x}_2 = m_2 - c_2 \phi x_1, \tag{15}$$

where c_2 may be regarded as a measure of effectiveness of player I's weapons against player II's defenses. By reversing the role of the players, we obtain the second state equation

$$\dot{x}_1 = m_1 - c_1 \psi x_2. \tag{16}$$

Suppose we plan on the war lasting some definite time T. Each day (say) player I puts $(1 - \phi)x_1$ into the field, and II puts in $(1 - \psi)x_2$. The difference

$$(1 - \psi)x_2 - (1 - \phi)x_1$$

reflects the margin of superiority for that day. If we integrate this over the time period $[0, T]$, we can measure the *accumulated margin of superiority*

$$C(\phi, \psi) = \int_0^T [(1 - \psi)x_2 - (1 - \phi)x_1]\, dt. \tag{17}$$

Player II's aim will be to choose ψ to *maximize* this quantity, and player I will choose ϕ to *minimize* it (i.e., to make it as largely negative as possible). We also suppose (without loss of generality) that $c_1 > c_2$. Otherwise we simply reverse the roles of the players.

In this formulation we have a fixed-time free-end-point differential game, and we can apply the HJB equation (12), that is, if $V(\mathbf{x}, t)$ represents the value of the game starting from $\mathbf{x}$ at time t, then

$$\frac{\partial V}{\partial t} = -\min_{0 \le \phi \le 1} \max_{0 \le \psi \le 1} \left\{ (m_1 - c_1 \psi x_2) \frac{\partial V}{\partial x_1} + (m_2 - c_2 \phi x_1) \frac{\partial V}{\partial x_2} + (1 - \psi) x_2 - (1 - \phi) x_1 \right\}, \qquad V(x, T) = 0, \quad x \in R^2. \tag{18}$$

To find the optimal controls ϕ^* and ψ^*, we just have to locate where the $\min_\phi \max_\psi$ occurs, or where the min max of

$$\psi \left\{ -x_2 \left(1 + c_1 \frac{\partial V}{\partial x_1} \right) \right\} + \phi \left\{ x_1 \left(1 - c_2 \frac{\partial V}{\partial x_2} \right) \right\}$$

occurs. Setting

$$S_1 = 1 - c_2 V_{x_2}, \qquad S_2 = -1 - c_1 V_{x_1} \tag{19}$$

in the case $x_1, x_2 > 0$, we have

$$\phi^* = \begin{cases} 0 & \text{if } S_1 > 0 \\ 1 & \text{if } S_1 < 0, \end{cases} \qquad \psi^* = \begin{cases} 0 & \text{if } S_2 < 0 \\ 1 & \text{if } S_2 > 0. \end{cases} \tag{20}$$

We shall return to the case $x_1(t) = 0$ or $x_2(t) = 0$ for some $t \in [0, T]$ at the end of the example.

By (14) along the optimal trajectory, we have

$$\dot{V}_{x_1} = -(S_1 \phi^* - 1), \qquad \dot{V}_{x_2} = -(S_2 \psi^* + 1), \qquad V(\mathbf{x}, T) = 0, \qquad x \in R^2. \tag{21}$$

We can simplify the analysis by substituting (21) into (19), arriving at

$$\dot{S}_1 = +c_2(S_2 \psi^* + 1), \qquad \dot{S}_2 = +c_1(S_1 \phi^* - 1) \tag{22}$$

with $S_1(T) = 1$, $S_2(T) = -1$.

The solution of this differential game follows in the usual way we solved the fixed-time control problem, by using the information given at time $t = T$ to compute the final control values, and then solving backward in time.

By Eqs. (22) and (20),

$$\phi^*(T) = 0, \qquad \psi^*(T) = 0,$$

so the game ends with both sides attacking fully. Hence, over the final time period

$$S_1(t) = 1 + c_2(t - T), \qquad S_2(t) = -1 + c_1(T - t).$$

The switching equations (20) tell us that the controls will maintain their value at 0 until either S_1 or S_2 changes sign. Simple calculation shows that S_1 changes sign at $T - 1/c_2$ and S_2 at $T - 1/c_1$. Since $c_1 > c_2$, $(T - 1/c_1) > (T - 1/c_2)$ so S_2 changes sign last. Hence

$$\psi^*(t) = \begin{cases} 0 & T - 1/c_1 \le t \le T \\ 1 & \quad ? < t < T - 1/c_1 \end{cases} \tag{23}$$

or ψ^* has its final switch at $T - 1/c_1$.

Next we continue solving backward from $T - 1/c_1$. Now

$$\psi^* = 1, \qquad \phi^* = 0. \tag{24}$$

Since S is continuous, we know

$$S_1\left(T - \frac{1}{c_1}\right) = 1 - \frac{c_2}{c_1}, \qquad S_2\left(T - \frac{1}{c_1}\right) = -1 + \frac{c_2}{c_1} \tag{25}$$

and, with control values (23), Eqs. (22) become

$$\dot{S}_1 = c_2(S_2 + 1), \qquad \dot{S}_2 = -c_1.$$

Solving backward, using (25), then gives

$$\dot{S}_1 = c_1 c_2(T - t), \qquad S_2 = -1 + c_1(T - t), \qquad t < T - 1/c_1,$$

$$S_1(t) = 1 - \frac{c_2}{2c_1} - \frac{c_1 c_2}{2}(t - T)^2, \qquad t \le T - 1/c_1, \tag{26}$$

$$S_2(t) = -1 + c_1(T - t), \qquad t \le T - 1/c_1. \tag{27}$$

It is easy to see that $S_2 > 0$ now, and so the switch can occur only when $S_1(t) = 0$. This happens at

$$t = T - \frac{1}{c_1}\sqrt{\frac{2c_1}{c_2} - 1};$$

as a consequence, ϕ^* switches from 0 to 1 at $T - (1/c_1)\sqrt{(2c_1/c_2) - 1}$. If we continue solving backward from $T - (1/c_1)\sqrt{(2c_1/c_2) - 1}$ with

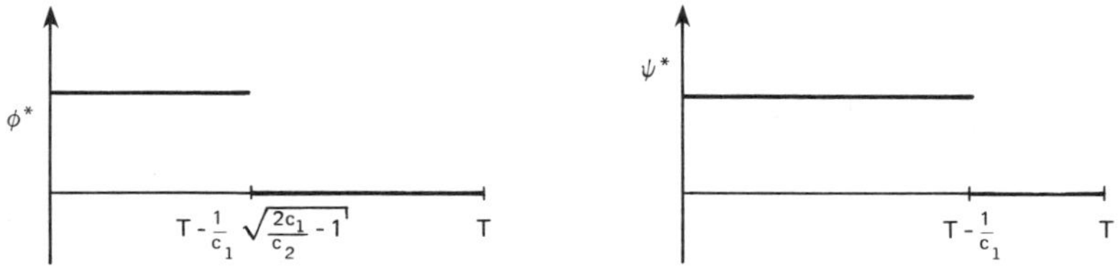

Fig. 7

$\phi^* = \psi^* = 1$, we shall find that no further switches occur, so ϕ^* and ψ^* must be the unique solutions of the HJB equation (see Fig. 7).

Now that we have explicitly calculated the optimal controls we can solve the state equations from any initial point $\mathbf{x}$, and calculate V:

$$V(x_1, x_2, 0) = C(\phi^*, \psi^*)$$
$$= \int_{T-(1/c_1)\sqrt{(2c_1/c_2)-1}}^{T-1/c_1} (-x_1^*(t))\,dt + \int_{T-1/c_1}^{T} (x_2^*(t) - x_1^*(t))\,dt$$

when $x_1^*(0) = x_1$, $x_2^*(0) = x_2$. Solving the state equations and substituting for the optimal trajectory gives

$$V(x_1, x_2, 0) = \frac{x_2}{2c_1} - \frac{m_2}{6c_1^2} + \left(\frac{m_2}{2c_2} - x_1\right)T$$
$$+ \frac{1}{2}(c_1 x_2 - m_1)T^2 - \frac{c_1 m_2 T^3}{6}.$$

Finally, we consider the case in which one of the $\mathbf{x}^*$, say x_1^*, becomes zero on $[0, T]$. We can see easily that this can happen only if $t \le T - 1/c_1$: for if $t \in (T - 1/c_1, T)$, then $\dot{x}_1^* = m_1 > 0$, and so x_1^* is increasing over $(T - 1/c_1, T)$. However, it is certainly possible that $x_1^*(t) = 0$ for $t \le T - 1/c_1$. This corresponds to player I being annihilated. There will be a critical trajectory that just hits the point A in Fig. 8. All trajectories which start off below this will have $x_1^*(t_0) = 0$ for some $t_0 < T - 1/c_1$. It is clear that if $x_1^*(t_0) = 0$, all player II must do is keep $\dot{x}_1^* = 0$; x_1^* can be kept at 0 only if

$$x_2^* \ge m_1/c_1, \tag{28}$$

and in this case II should play

$$\psi^*(t) = m_1/c_1 x_2(t), \qquad t_0 \le t \le T - 1/c_1,$$

until point A, when both ϕ^* and ψ^* revert to their old optimum strategies $(+1)$.

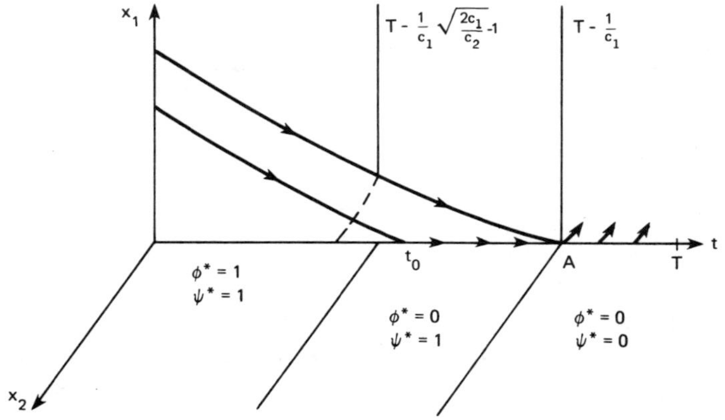

Fig. 8

Finally, for $t \leq t_0$, we can resolve the HJB equation with $x_1(t_0) = 0$, and we find that II's optimal strategy is unchanged; however, player I should switch slightly earlier, along the dotted line in Fig. 8. Here we see that since I's final attack is going to be nullified through the annihilation of his forces, he tries to compensate by starting earlier. Of course, the value of the game will have to be recomputed for these new optimum trajectories. The case in which II is annihilated can be handled symmetrically.

Problems

1. Find the feedback control and minimal cost for the infinite horizon control problem

$$\dot{x} = -x + u, \qquad x(0) = x_0,$$

with cost

$$C(u) = \int_0^\infty [(x(t))^2 + (u(t))^2]\, dt.$$

2. Find the control $\mathbf{u} = (u_1, u_2)$ that minimizes the cost

$$\frac{1}{2}\int_0^\infty (x^2 + u_1^2 + u_2^2)\, dt$$

for the system

$$\dot{x} = u_1 + u_2, \qquad x(0) = 1.$$

3. Calculate the feedback control that minimizes

$$\frac{1}{2}\int_0^T ((x_1 - \alpha)^2 + u^2)\,dt$$

for the system

$$\dot{x}_1 = x_2, \qquad \dot{x}_2 = u, \qquad \mathbf{x}(0) = \mathbf{x}^0.$$

4. Calculate the symmetric feedback matrix

$$E = \begin{bmatrix} e_1 & e \\ e & e_2 \end{bmatrix}$$

for the optimal control of the system

$$\ddot{x} - x = u,$$

with cost

$$C(u) = \int_0^\infty (\mathbf{x}^{\mathrm{T}} W \mathbf{x} + \gamma u^2)\,dt,$$

where

$$x = \begin{bmatrix} x \\ \dot{x} \end{bmatrix}, \qquad W = \begin{bmatrix} w_1 & w \\ w & w_2 \end{bmatrix} > 0, \qquad \gamma > 0.$$

5. By use of dynamic programming, derive the optimality conditions for the discrete control problem

$$\text{minimize } \frac{1}{2}\sum_{k=0}^{9} u_k^2$$

subject to

$$x_{k+1} = x_k + \alpha u_k, \qquad x_0 = 1.$$

References

[1] A. Friedman, "Differential Games." Wiley, Interscience, New York, 1971.
[2] R. Isaacs, "Differential Games." Wiley, New York, 1965.
[3] E. Lee and L. Markus, "Foundations of Optimal Control Theory." Wiley, New York, 1967.

Chapter VII Controllability and Observability

One of our central assumptions in the time-optimal control problem was that the target point could be reached in finite time by an admissible control. In this chapter we investigate this problem of *controllability* of systems.

1. CONTROLLABLE LINEAR SYSTEMS

To begin with, suppose that there are no restrictions on the magnitude of the control $\mathbf{u}(t)$. Let the restraint set be $\Omega = R^m$.

Definition 1 The linear control process

$$\dot{\mathbf{x}}(t) = A(t)\mathbf{x}(t) + B(t)\mathbf{u}(t), \tag{1}$$

with $\Omega = R^m$, is called (completely) controllable at t_0 if to each pair of points $\mathbf{x}_0$ and $\mathbf{x}_1$ in R^n there exists a piecewise continuous control $\mathbf{u}(t)$ on some finite interval $t_0 \leq t \leq t_1$, which steers $\mathbf{x}_0$ to $\mathbf{x}_1$. In the case (1) is autonomous, we shall see that controllability does not depend upon the initial time, and so we just say the system is controllable.

Theorem 1 The autonomous linear process

$$\dot{\mathbf{x}} = A\mathbf{x} + B\mathbf{u} \tag{2}$$

is controllable if and only if the $n \times nm$ matrix

$$[B, AB, A^2B, \ldots, A^{n-1}B]$$

has *rank n.*

Proof Assume that (2) is controllable, but that

$$\operatorname{rank}[B, AB, \ldots, A^{n-1}B] < n.$$

Consequently, we can find a vector $\mathbf{v} \in R^n$ such that

$$\mathbf{v}[B, AB, A^2B, \ldots, A^{n-1}B] = \mathbf{0}$$

or

$$\mathbf{v}B = \mathbf{v}AB = \cdots = \mathbf{v}A^{n-1}B = \mathbf{0}. \tag{3}$$

By the Cayley–Hamilton theorem [1], A satisfies its own characteristic equation, so for some real numbers $c_1, c_2, \ldots, c_n$,

$$A^n = c_1 A^{n-1} + c_2 A^{n-2} + \cdots + c_n I.$$

Thus

$$\mathbf{v}A^nB = c_1 \mathbf{v}A^{n-1}B + c_2 \mathbf{v}A^{n-2}B + \cdots + c_n \mathbf{v}B = \mathbf{0}$$

by (2). By induction, $\mathbf{v}A^{n+k}B = \mathbf{0}$ for all $k = 0, 1, 2, \ldots$, or $\mathbf{v}A^mB = \mathbf{0}$ for all $m = 0, 1, 2, \ldots$, and so,

$$\mathbf{v}e^{At}B = \mathbf{v}\left[I + At + \frac{1}{2!}A^2t^2 + \cdots\right]B = \mathbf{0} \tag{4}$$

for all real t. However, the response starting from $\mathbf{x}_0 = 0$ with control $\mathbf{u}$ is just

$$\mathbf{x}(t) = e^{At}\int_0^t e^{-As}B\mathbf{u}(s)\,ds,$$

so (4) implies that

$$\mathbf{v}\cdot\mathbf{x}(t) = \int_0^t \mathbf{v}e^{A(t-s)}B\mathbf{u}(s)\,ds = 0,$$

for all $\mathbf{u}$ and $t > 0$. In other words, all points reachable from $\mathbf{x}_0 = \mathbf{0}$ are orthogonal to $\mathbf{v}$, which contradicts the definition of controllability.

Next suppose that $\operatorname{rank}[B, AB, A^2B, \ldots, A^{n-1}B] = n$. Note first that (1) is controllable if and only if

$$\bigcup_{t>0}\left\{\int_0^t e^{A(t-s)}B\mathbf{u}(s)\,ds : \mathbf{u} \text{ is piecewise continuous}\right\} = R^n. \tag{5}$$

Suppose (4) is false; then certainly

$$\left\{\int_0^1 e^{A(1-s)}B\mathbf{u}(s)\,ds : \mathbf{u} \text{ is piecewise continuous}\right\} \neq R^n$$

or there must exist some nonzero $\mathbf{v} \in R^n$ such that

$$\mathbf{v}\int_0^1 e^{A(1-s)}B\mathbf{u}(s)\,ds = \mathbf{0}$$

for all piecewise continuous $\mathbf{u}$, and so

$$\mathbf{v}e^{A(1-s)}B = \mathbf{0}, \qquad 0 \leq s \leq 1. \tag{6}$$

Setting $s = 1$, we see that $\mathbf{v}B = \mathbf{0}$. Differentiating (6) and setting $s = 1$ give $\mathbf{v}AB = \mathbf{0}$. Accordingly, by induction, $\mathbf{v}B = \mathbf{v}Ab = \mathbf{v}A^2B = \cdots = \mathbf{v}A^{n-1}B = \mathbf{0}$, which contradicts the rank condition. ■

Corollary 1 If (1) is normal, then it is controllable.

The converse is true only in the case $m = 1$; in fact, for $m = 2$, we have a simple counter example.

Example 1

$$\begin{bmatrix}\dot{x}_1\\ x_2\end{bmatrix} = \begin{bmatrix}-1 & 0\\ 0 & -2\end{bmatrix}\begin{bmatrix}x_1\\ x_2\end{bmatrix} + \begin{bmatrix}1 & 0\\ 1 & 1\end{bmatrix}\begin{bmatrix}u_1\\ u_2\end{bmatrix},$$

$$A = \begin{bmatrix}-1 & 0\\ 0 & -2\end{bmatrix}, \qquad \mathbf{b}^1 = \begin{bmatrix}1\\ 1\end{bmatrix}, \qquad \mathbf{b}^2 = \begin{bmatrix}0\\ 1\end{bmatrix},$$

$$A\mathbf{b}^1 = \begin{bmatrix}-1\\ -2\end{bmatrix}, \qquad A\mathbf{b}^2 = \begin{bmatrix}0\\ -2\end{bmatrix}.$$

Hence $\mathbf{b}^2$ and $A\mathbf{b}^2$ are linearly dependent, so system is not normal. However

$$[B, AB] = \begin{bmatrix}1 & 0 & -1 & 0\\ 1 & 1 & -2 & -2\end{bmatrix}$$

is certainly rank 2.

Note also that the proof of Theorem 1 shows that if the system is controllable, then it is controllable in any arbitrarily small time.

The concept of controllability for autonomous systems is independent of the coordinate system chosen for R^n. For if $\mathbf{y} = \mathrm{Q}\mathbf{x}$, where Q is a real nonsingular constant matrix, and

$$\mathscr{A} = QAQ^{-1}, \qquad \mathscr{B} = QB,$$

then

$$\dot{\mathbf{x}} = A\mathbf{x} + B\mathbf{u} \tag{7}$$

is controllable if and only if

$$\dot{\mathbf{y}} = \mathscr{A}\mathbf{y} + \mathscr{B}\mathbf{u} \tag{8}$$

is controllable. This follows easily from the identity

$$\begin{aligned}\operatorname{rank}[B, AB, \ldots, A^{n-1}B] &= \operatorname{rank} Q[B, AB, \ldots, A^{n-1}B] \\ &= \operatorname{rank}[\mathscr{B}, \mathscr{A}\mathscr{B}, \ldots, \mathscr{A}^{n-1}\mathscr{B}].\end{aligned}$$

We shall say (7) and (8) are *equivalent control systems.*

Example 2 The autonomous linear process

$$y^{(n)} + a_1 y^{(n-1)} + a_2 y^{(n-2)} + \cdots + a_n y = u \tag{L}$$

is controllable with $u \in R$. In phase space, (L) is just

$$\dot{\mathbf{y}} = \begin{bmatrix} 0 & 1 & 0 & \cdots & 0 \\ 0 & 0 & 1 & \cdots & 0 \\ \vdots & \vdots & \vdots & & \vdots \\ 0 & 0 & 0 & \cdots & 1 \\ -a_n & -a_{n-1} & -a_{n-2} & \cdots & -a_1 \end{bmatrix} \mathbf{y} + \begin{bmatrix} 0 \\ 0 \\ 0 \\ \vdots \\ 1 \end{bmatrix} u,$$

and the rank condition is easily checked.

For autonomous systems with one control variable, this example is typical.

Theorem 2 Every autonomous controllable system

$$\dot{\mathbf{x}} = A\mathbf{x} + \mathbf{b}u, \tag{9}$$

with $u \in R$, is equivalent to a system of the form (L).

Proof Define the real nonsingular (by Theorem 1) $n \times n$ matrix

$$Q = [A^{n-1}\mathbf{b}, A^{n-2}\mathbf{b}, \ldots, A\mathbf{b}, \mathbf{b}].$$

Direct calculation shows that

$$\mathbf{b} = Q\mathbf{d} \qquad \text{and} \qquad AQ = Q\mathscr{A},$$

where

$$\mathscr{A} = \begin{bmatrix} \alpha_1 & 1 & 0 & \cdots & 0 \\ \alpha_2 & 0 & 1 & \cdots & 0 \\ \vdots & \vdots & \vdots & & \vdots \\ \alpha_{n-1} & 0 & 0 & \cdots & 1 \\ \alpha_n & 0 & 0 & \cdots & 0 \end{bmatrix}, \qquad \mathbf{d} = \begin{bmatrix} 0 \\ 0 \\ \vdots \\ 0 \\ 1 \end{bmatrix},$$

and the real constants $\alpha_1, \alpha_2, \ldots, \alpha_n$ are determined by the characteristic equation for A by

$$A^n = \alpha_1 A^{n-1} + \alpha_2 A^{n-2} + \cdots + \alpha_n I. \tag{10}$$

The analogous change of coordinates applied to (L) would give system matrices $\mathscr{A}'$ and $\mathbf{d}$, where

$$\mathscr{A}' = \begin{bmatrix} -a_1 & 1 & 0 & \cdots & 0 \\ -a_2 & 0 & 1 & \cdots & 0 \\ \vdots & \vdots & \vdots & & \vdots \\ -a_{n-1} & 0 & 0 & \cdots & 1 \\ -a_n & 0 & 0 & \cdots & 0 \end{bmatrix}.$$

Consequently, (9) will be equivalent to (L) if we choose

$$\alpha_1 = -a_1, \quad \alpha_2 = -a_2, \quad \ldots, \quad \alpha_n = -a_n.$$

In other words, (9) is equivalent to the system

$$y^{(n)} - \alpha_1 y^{(n-1)} - \alpha_2 y^{(n-2)} - \cdots - \alpha_n y = u,$$

where $\alpha_1, \ldots, \alpha_n$ are obtained from the characteristic equation for A by (10). ■

The next concept, which occurs often, is the *domain* of *null controllability at time* t_0. This is the set of all points in R^n that can be steered to the origin in time $t_1 > t_0$ by an admissible controller $\mathbf{u}(t)$ taking values in a given restraint set $\Omega \subset R^m$. In the case $\Omega = R^m$ and the system is autonomous, it is easy to show that null controllability is equivalent to controllability (Problem 2). In Section II.3 we showed directly, by constructing time optimal controllers, that certain second-order systems were null controllable with controls restricted by $|u(t)| \leq 1$. The following theorem essentially generalizes this construction.

Theorem 3 (Hermes and LaSalle [2]) Consider the autonomous linear process in R^n

$$\dot{\mathbf{x}} = A\mathbf{x} + B\mathbf{u}, \tag{11}$$

for which

(a) $\mathbf{0}$ belongs to the interior of Ω;
(b) $\operatorname{rank}[B, AB, A^{n-1}B] = n$;
(c) every eigenvalue λ of A satisfies $\operatorname{Re}(\lambda) \leq 0$.

Then the domain of null controllability at any time T equals R^n.

Proof Without loss of generality, assume that Ω contains the unit cube in R^n. Then the theorem will be proven if we can show that

$$\mathscr{R} = U_{t \geq 0}\mathscr{R}(t) = \{\mathbf{y}(t:\mathbf{u}) : |u_i| \leq 1, t > 0\} = R^n,$$

where $\mathscr{R}(t)$ is the reachable set introduced in Appendix 1, and

$$\mathbf{y}(t:\mathbf{u}) = \int_0^t e^{-A\tau}B\mathbf{u}(\tau)\, d\tau.$$

Since $\mathscr{R}$ is convex, it is enough to show that for each $\boldsymbol{\eta} \neq \mathbf{0}$ in R^n, there exists an admissible control $\mathbf{u}^*$ such that $\boldsymbol{\eta} \cdot \mathbf{y}(t:\mathbf{u}^*) \to \infty$ as $t \to \infty$. (Geometrically this is clear; a detailed proof can be obtained by supposing $\mathscr{R} \neq R^n$, then using the supporting hyperplane theorem to derive a contradiction.) We show that the admissible control

$$\mathbf{u}^*(t) = \operatorname{sgn}[\boldsymbol{\eta}^{\mathrm{T}} e^{-At} B], \qquad t > 0,$$

suffices. Note first that the rank condition implies that at least one component of $\boldsymbol{\eta}^{\mathrm{T}} e^{-At} B$, say $\boldsymbol{\eta}^{\mathrm{T}} e^{-At}\mathbf{b}_1$, is not identically zero. Then

$$\boldsymbol{\eta}^{\mathrm{T}}\mathbf{y}(t:\mathbf{u}^*) = \int_0^t |\boldsymbol{\eta}^{\mathrm{T}} e^{-A\tau} B\mathbf{u}^*(\tau)|\, d\tau,$$

and hence $\boldsymbol{\eta}^{\mathrm{T}}\mathbf{y}(t:\mathbf{u}^*) \to \infty$, if we can show that

$$\int_0^\infty |\boldsymbol{\eta}^{\mathrm{T}} e^{-A\tau} B|\, d\tau = \infty. \tag{12}$$

Suppose (12) is false and set

$$v(t) = \int_t^\infty [\boldsymbol{\eta}^{\mathrm{T}} e^{-A\tau}\mathbf{b}_1]\, d\tau. \tag{13}$$

If $\varphi(\lambda)$ is the characteristic polynomial of A, then $\varphi(-\lambda)$ is the characteristic polynomial for $-A$, and hence

$$\varphi(-D)[\boldsymbol{\eta}^{\mathrm{T}} e^{-At}\mathbf{b}_1] = 0, \qquad 0 < t < \infty,$$

where $D \equiv d/dt$. Consequently, from (13)

$$(\varphi(-D)Dv)(t) = (D\varphi(-D)v)(t) = 0, \qquad 0 < t < \infty. \tag{14}$$

However, the characteristic polynomial for the differential operator $D\varphi(-D)$ is just $\lambda\varphi(-\lambda)$, and by assumption (c), its roots must all have nonnegative real parts. Hence the solution v of the homogeneous equation (10) cannot decay to 0 as $t \to \infty$, and so

$$\int_0^\infty |\boldsymbol{\eta}^T e^{-A\tau}\mathbf{b}_1|\, d\tau = \infty,$$

as required. ■

One of the central concerns of classical control theory is to design feedback controllers (preferably linear) that will stabilize a given system, and hence improve its sensitivity to noise, measurement errors, etc. The question of the existence of a linear feedback control $\mathbf{u} = K\mathbf{x}$, for which the closed-loop system

$$\dot{\mathbf{x}} = (A + BK)\mathbf{x}$$

is stable, is intimately related to the concept of controllability. Namely, we say the control system (2) is $(+)$-*stabilizable* if there exists an $m \times n$ constant matrix K^+ such that

$$A^+ = A + BK^+$$

has eigenvalues with only negative real parts. The same system is $(-)$-*stabilizable* if there exists a matrix K^- such that

$$A^- = A + BK^-$$

has eigenvalues with only positive real parts.

Theorem 4 (Russell [3]) The autonmous system (2) is controllable if and only if it is both $(+)$- and $(-)$-stabilizable.

Proof Suppose (2) is $(+)$- and $(-)$-stabilizable. As remarked earlier, it is enough to show that (2) is null controllable in some time T. Set $\mathbf{x}_0 \in R^n$, and define

$$\begin{aligned} &\dot{\mathbf{x}}^+(t) = A^+\mathbf{x}^+(t), && \mathbf{x}^+(0) = \mathbf{x}_0, \\ &\dot{\mathbf{x}}^-(t) = A^-\mathbf{x}^-(t), && \mathbf{x}^-(T) = -\mathbf{x}^+(T), \\ &\mathbf{u}^+(t) = K^+\mathbf{x}^+(t), && \\ &\mathbf{u}^-(t) = K^-\mathbf{x}^-(t), && 0 \le t \le T. \end{aligned}$$

These equations have solutions

$$\mathbf{x}^+(t) = e^{A^+t}\mathbf{x}_0$$

and

$$\mathbf{x}^-(t) = -e^{-A^-(T-t)}\mathbf{x}^+(T) = -e^{-A^-(T-t)}e^{A^+T}\mathbf{x}_0.$$

Finally, define $\mathbf{f}_T : R^n \to R^n$, by

$$\mathbf{f}_T(\mathbf{x}_0) = \mathbf{x}^-(0) = -e^{-A^-T}e^{A^+T}\mathbf{x}_0.$$

Then

$$\mathbf{x}(t) = \mathbf{x}^+(t) + \mathbf{x}^-(t) \qquad \text{and} \qquad \mathbf{u}(t) = \mathbf{u}^+(t) + \mathbf{u}^-(t)$$

satisfy

$$\dot{\mathbf{x}} = A\mathbf{x} + B\mathbf{u}, \qquad \mathbf{x}(0) = \mathbf{x}_0 + \mathbf{f}_T(\mathbf{x}_0), \qquad \mathbf{x}(T) = \mathbf{0},$$

or the control $\mathbf{u}$ steers $\mathbf{x}_0 + \mathbf{f}_T(\mathbf{x}_0)$ to $\mathbf{0}$ in time T. If we can show that for some $T > 0$, $\{\mathbf{x}_0 + \mathbf{f}_T(\mathbf{x}_0)\}$ ranges over R^n as $\mathbf{x}_0 \in R^n$, (2) must be null controllable in that time T, and the result follows. However,

$$\mathbf{x}_0 + \mathbf{f}_T(\mathbf{x}_0) = (I - e^{-A^-T}e^{A^+T})\mathbf{x}_0,$$

and by the $(+)$ and $(-)$ stability assumptions we can guarantee that

$$\|e^{-A^-T}e^{A^+T}\| < 1$$

by choosing T sufficiently large. For such a T, the mapping $(I + \mathbf{f}_T)$: $R^n \to R^n$ is 1–1 and onto (since $\mathbf{f}_T$ is a contraction), and hence $\{\mathbf{x}_0 + \mathbf{f}_T(\mathbf{x}_0) : \mathbf{x}_0 \in R^n\} = R^n$ as required.

To prove the converse, we take $m = 1$ for simplicity. Note that if $A + BK$ is a stability matrix, then so is $Q(A + BK)Q^{-1} = \mathscr{A} + \mathscr{B}\mathscr{K}$ $(\mathscr{K} = KQ^{-1})$. Accordingly, by Theorem 2 it is sufficient to show that the system equivalent to (2)

$$y^{(n)} - \alpha_1 y^{(n-1)} - \alpha_2 y^{(n-2)} - \cdots - \alpha^n y = u,$$

is $(+)$- and $(-)$-stabilizable. For this problem, the linear feedback control

$$u = d_n y_1 + d_{n-1} y_2 + \cdots + d_1 y_n, \qquad \mathbf{d} = (d_1, d_2, \ldots, d_n) \in R^n,$$

gives the closed-loop system

$$y^{(n)} - (\alpha_1 + d_1)y^{(n-1)} - (\alpha_2 + d_2)y^{(n-2)} - \cdots - (\alpha_n + d_n)y = 0, \quad (15)$$

and a $(\pm)$ stable process can be achieved by determining $\mathbf{d}$ so that the characteristic polynomial of (15) is

$$\lambda^n - (\alpha_1 + d_1)\lambda^{n-1} - \cdots - (\alpha_n + d_n) = (\lambda \pm 1)^n$$

or

$$-(\alpha_i + d_i) = \binom{n}{i}(\pm 1)^i, \qquad 1 \le i \le n. \quad \blacksquare \tag{16}$$

With a little more work the above proof shows that if (2) is controllable, then by a suitable choice of the feedback matrix K, one place the eigenvalues of $A + BK$ anywhere in the left-hand plane. Such a K can be constructed by reversing the proof.

Finally, we give a simple sufficient condition for the nonautonomous system (1) to be controllable. Suppose the matrix $A(t)$ is $k - 1$ times differentiable and $B(t)$ is $k - 2$ times differentiable for $t \ge t_0$. Define inductively a sequence of $n \times m$ matrices $M_j(t)$ as the solutions of the equation

$$M_{j+1}(t) = -A(t)M_j(t) + \frac{d}{dt}(M_j(t)), \qquad j = 0, 1, 2, \ldots, k-2,$$

with

$$M_0(t) = B(t), \qquad t \ge t_0.$$

Then

$$\begin{aligned}\frac{d}{dt}(X^{-1}(t)B(t)) &= X^{-1}(t)[-A(t)B(t) + \frac{d}{dt}B(t)]\\ &= X^{-1}(t)M_1(t)\end{aligned}$$

and similarly

$$\frac{d^j}{dt^j}(X^{-1}(t)B(t)) = X^{-1}(t)M_j(t), \qquad t \ge t_0, \quad j = 0, 1, 2, \ldots, k-1. \tag{17}$$

Theorem 5 (Hermes and LaSalle [2]) If $A(t)$ has $k - 2$ derivatives and $B(t)$ has $k - 1$ derivatives for $t \ge t_0$, and if for some positive integer k and each $t_1 > t_0$, there exists a $t \in [t_0, t_1)$ such that

$$\operatorname{rank}[M_0(t), M_1(t), \ldots, M_{k-1}(t)] = n,$$

then (1) is controllable at t_0.

Proof Following the proof of Theorem 1, it can be shown by using the continuity of $X^{-1}B$ that if (1) is not controllable, then there exists a $\boldsymbol{\eta} \ne \mathbf{0}$ and a $t_1 > t_0$ such that

$$\boldsymbol{\eta}^{\mathrm{T}} X^{-1}(t)B(t) = \mathbf{0}, \qquad t_0 \le t < t_1. \tag{18}$$

Differentiating (18) and applying (17) gives

$$\boldsymbol{\eta}^{\mathrm{T}}X^{-1}(t)M_0(t)=\mathbf{0}, \qquad \boldsymbol{\eta}^{\mathrm{T}}X^{-1}(t)M_1(t)=\mathbf{0} \quad \ldots, \quad \boldsymbol{\eta}^{\mathrm{T}}X^{-1}(t)M_{k-1}(t)=\mathbf{0},$$

for $t \in [t_0, t_1)$. Consequently $\boldsymbol{\eta}^{\mathrm{T}}X^{-1}(t)$ is nonzero and is orthogonal to $[M_0(t), \ldots, M_{k-1}(t)]$, which contradicts the rank condition. ■

Example 3 Show that the system

$$\dot{\mathbf{x}} = \begin{bmatrix} t^3 & 0 \\ 0 & t^4 \end{bmatrix} \mathbf{x} + \begin{bmatrix} t^2 \\ t^3 \end{bmatrix} u$$

is controllable at $t_0 = 0$.

Computation shows that

$$M_0(0) = \mathbf{b}(0) = \begin{bmatrix} 0 \\ 0 \end{bmatrix},$$

$$M_1(0) = -A(0)\mathbf{b}(0) + \dot{\mathbf{b}}(0) = \begin{bmatrix} 0 \\ 0 \end{bmatrix},$$

and

$$\dot{M}_1(0) = -A\dot{\mathbf{b}}(0) - \dot{A}(0)\mathbf{b}(0) + \ddot{\mathbf{b}}(0) = \begin{bmatrix} 2 \\ 0 \end{bmatrix}.$$

Hence

$$M_2(0) = A(0)M_1(0) + \dot{M}_1(0) = \begin{bmatrix} 2 \\ 0 \end{bmatrix}.$$

Similarly, it can be shown that

$$M_3(0) = \begin{bmatrix} 0 \\ 6 \end{bmatrix},$$

and so $\{M_2(0), M_3(0)\}$ are linearly independent.

2. OBSERVABILITY

Suppose that we can observe some known linear function

$$\mathbf{y}(t) = C\mathbf{x}(t), \tag{1}$$

where C is an $m \times n$ matrix of the output $\mathbf{x}(t)$ of a known dynamical system

$$\dot{\mathbf{x}} = A\mathbf{x} \tag{2}$$

whose initial condition $\mathbf{x}(0) = \mathbf{x}_0$, however, is unknown. For instance, if

$$C = \begin{bmatrix} 1 & 0 & \cdots & 0 \\ 0 & 1 & \cdots & 0 \\ 0 & 0 & \cdots & 0 \\ \vdots & \vdots & \vdots & \vdots \\ 0 & 0 & \cdots & 0 \end{bmatrix},$$

then (1) would mean that we would observe the first two components of $\mathbf{x}(t)$. Given such a framework the natural question to ask is, How much of $\mathbf{x}$ need we observe in order to be able to determine $\mathbf{x}$ uniquely? Since $\mathbf{x}$ is completely determined by (2) once the initial condition $\mathbf{x}_0$ is known, we call the system (1)–(2) *observable* if the knowledge of $\mathbf{y}(t)$ on some interval $[0, T]$ uniquely determines $\mathbf{x}_0$. Of course once we know a system is observable, the next problem is to find a recipe for reconstructing $\mathbf{x}_0$ from $\{\mathbf{y}(t)\}$.

It turns out the solution of both these problems is intimately related to the controllability concept introduced earlier.

Theorem 1 The system (1)–(2) is observable if and only if the control system

$$\dot{\mathbf{z}} = A^{\mathrm{T}}\mathbf{z} + C^{\mathrm{T}}\mathbf{u} \tag{3}$$

is controllable. Consequently (1)–(2) is observable if and only if

$$\operatorname{rank}[C^{\mathrm{T}}, A^{\mathrm{T}}C^{\mathrm{T}}, (A^{\mathrm{T}})^2C^{\mathrm{T}}, \ldots, (A^{\mathrm{T}})^{n-1}C^{\mathrm{T}}] = n. \tag{4}$$

Proof We can write the solution of (2)

$$\mathbf{x}(t) = e^{At}\mathbf{x}_0, \qquad t > 0,$$

and so, by (1),

$$\mathbf{y}(t) = Ce^{At}\mathbf{x}_0, \qquad t > 0. \tag{5}$$

By definition the system is observable at any time $T > 0$ if and only if there does not exist an $\mathbf{x}_0 \neq 0$ for which

$$\mathbf{y}(t) = Ce^{At}\mathbf{x}_0 \equiv \mathbf{0}, \qquad 0 \le t \le T. \tag{6}$$

Suppose that such an $\mathbf{x}_0$ exists. Then setting $t = 0$ in (6) gives $C\mathbf{x}_0 = \mathbf{0}$. Differentiating (6) and setting $t = 0$ gives $CA\mathbf{x}_0 = 0$, etc. So by induction we see that

$$C\mathbf{x}_0 = CA\mathbf{x}_0 = CA^2\mathbf{x}_0 = \cdots = CA^{n-1}\mathbf{x}_0 = \mathbf{0},$$

that is,

$$\mathbf{x}_0^{\mathrm{T}}C^{\mathrm{T}} = \mathbf{x}_0^{\mathrm{T}}A^{\mathrm{T}}C^{\mathrm{T}} = \cdots = \mathbf{x}_0^{\mathrm{T}}(A^{\mathrm{T}})^{n-1}C^{\mathrm{T}} = \mathbf{0}$$

and the rank condition (4) fails. The converse is proven using the Cayley–Hamilton theorem in essentially the same way as before. ■

For the problem of calculating $\mathbf{x}_0$, note that, by (5)

$$\int_0^T (e^{At})^{\mathrm{T}}C^{\mathrm{T}}\mathbf{y}(t)\,dt = \int_0^T (e^{At})^{\mathrm{T}}C^{\mathrm{T}}Ce^{At}\mathbf{x}_0\,dt = M\mathbf{x}_0,$$

where

$$M = \int_0^T (e^{At})^{\mathrm{T}}C^{\mathrm{T}}Ce^{At}\,dt. \tag{7}$$

If the system is not observable by (6), there exists an $\mathbf{x}_0 \neq 0$ for which $M\mathbf{x}_0 = 0$, and hence M is singular. Conversely, if M is nonsingular, then

$$\mathbf{x}_0 = M^{-1}\int_0^T (e^{At})^{\mathrm{T}}C^{\mathrm{T}}\mathbf{y}(t)\,dt \tag{8}$$

is completely determined. So (1)–(2) is observable if and only if M is nonsingular, in which case $\mathbf{x}_0$ can be computed from (8). Notice that for autonomous systems, observability is independent of $T > 0$.

These results also carry over to the (perhaps from practical considerations more interesting) discrete time case of (1)–(2). For example, if

$$\mathbf{x}((k+1)\tau) = A\mathbf{x}(k\tau), \qquad \mathbf{x}(0) = \mathbf{x}_0, \tag{9}$$

$$\mathbf{y}(k\tau) = C\mathbf{x}(k\tau), \qquad k = 0, 1, 2, \ldots, \tag{10}$$

where τ represents some sampling period, then we say the system (9)–(10) is observable if $\mathbf{x}_0$ is uniquely determined once $\{\mathbf{y}(0), \mathbf{y}(\tau), \ldots, \mathbf{y}((n-1)\tau)\}$ are known.

Theorem 2 The system (9)–(10) is observable if and only if

$$\operatorname{rank}[C^{\mathrm{T}}, A^{\mathrm{T}}C^{\mathrm{T}}, \ldots, (A^{\mathrm{T}})^{n-1}C^{\mathrm{T}}] = n.$$

Analogous to (7)–(8) we can compute $\mathbf{x}_0$ from the observed $\mathbf{y}'^s$, as

$$\mathbf{x}(k\tau) = A^k\mathbf{x}_0, \qquad \mathbf{y}(k\tau) = CA^k\mathbf{x}_0.$$

Hence

$$\sum_{k=0}^{n-1} (A^k)^{\mathrm{T}}C^{\mathrm{T}}\mathbf{y}(k\tau) = \left(\sum_{k=0}^{n-1} (A^k)^{\mathrm{T}}C^{\mathrm{T}}CA^k\right)\mathbf{x}_0 = \bar{M}\mathbf{x}_0,$$

where $\bar{M}$ is the $n \times n$ matrix

$$\bar{M} = \sum_{k=0}^{n-1} (A^k)^T C^T C A^k.$$

Since $\bar{M}$ can alternatively be written

$$\bar{M} = KK^T$$

where K is the $n \times nm$ matrix

$$K = [C^T, A^T C^T, \ldots, (A^{n-1})^T C^T],$$

if the system is observable, then the rank condition guarantees M is nonsingular, and hence

$$\mathbf{x}_0 = \bar{M}^{-1} \sum_{k=0}^{n-1} (A^k)^T C^T \mathbf{y}(k\tau). \tag{11}$$

Clearly, any n-consecutive observations suffice in (11).

Problems

1. Check the controllability and observability of the following systems:

$$\dot{\mathbf{x}} = \begin{bmatrix} 0 & 1 \\ 0 & t \end{bmatrix} \mathbf{x} + \begin{bmatrix} 0 \\ 1 \end{bmatrix} u, \qquad y = [0 \quad 1]\mathbf{x};$$

$$\dot{\mathbf{x}} = \begin{bmatrix} 0 & 1 & 0 \\ 0 & 0 & 1 \\ -2 & -4 & -3 \end{bmatrix} \dot{\mathbf{x}} + \begin{bmatrix} 1 & 0 \\ 0 & 1 \\ -1 & 1 \end{bmatrix} \mathbf{u}, \qquad y = \begin{bmatrix} 0 & 1 & -1 \\ 1 & 2 & 1 \end{bmatrix} \mathbf{x}.$$

2. Show that the linear automonous control system is controllable if and only if it is null controllable.

3. Show that the nth-order controlled linear differential equation

$$x^{(n)}(t) + a_{n-1}(t)x^{(n-1)}(t) + \ldots + a_0(t)x(t) = u(t)$$

(or equivalently $\dot{\mathbf{x}}(t) = A(t)\mathbf{x}(t) + \mathbf{b}u(t)$,

$$A(t) = \begin{bmatrix} 0 & 1 & 0 & \ldots & 0 \\ 0 & 0 & 1 & \ldots & 0 \\ \vdots & \vdots & \vdots & & \vdots \\ 0 & 0 & 0 & \ldots & 1 \\ -a_0 & -a_1 & -a_2 & \ldots & -a_{n-1} \end{bmatrix}, \qquad \mathbf{b} = \begin{bmatrix} 0 \\ 0 \\ \vdots \\ 1 \end{bmatrix},$$

where a_j has j continuous derivatives), is controllable at any time $t_0 \geq 0$.

4. Let

$$A = \begin{bmatrix} 0 & -1 \\ 1 & 0 \end{bmatrix}, \qquad \mathbf{b}(t) = \begin{bmatrix} \cos t \\ \sin t \end{bmatrix}.$$

Shows that all solutions of $\dot{\mathbf{x}} = A\mathbf{x} + \mathbf{b}u$, $\mathbf{x}(0) = \mathbf{0}$, lie on the surface $x_1 \sin t - x_2 \cos t = 0$, and hence this system is not controllable at $t_0 = 0$.

5. Show that if A has two (or more) linearly independent eigenvectors corresponding to the same eigenvalue, $\dot{\mathbf{x}} = A\mathbf{x} + \mathbf{b}u$ cannot be controllable.

6. For the system

$$\dot{\mathbf{x}} = \begin{bmatrix} -1 & 1 & 0 \\ 0 & -1 & 0 \\ 0 & 0 & -2 \end{bmatrix} \mathbf{x} + \begin{bmatrix} 0 \\ 1 \\ 1 \end{bmatrix} u, \qquad \mathbf{v} = [1 \quad 1 \quad 1]\mathbf{x},$$

can the initial state $\mathbf{x}(0)$ be chosen so that the observed output $y(t) = te^{-t}$ for $t > 0$?

7. Show that the discrete linear autonomous control system

$$\mathbf{x}(k+1) = A\mathbf{x}(k) + B\mathbf{u}(k)$$

is controllable if and only if rank $[B, AB, A^2B, \ldots, A^{n-1}B] = n$.

8. Show that for the autonomous linear process

$$\dot{\mathbf{x}} = A\mathbf{x} + B\mathbf{u},$$

if there exists a nonzero m-vector $\mathbf{w}$ such that

$$B\mathbf{w}, \quad AB\mathbf{w}, \quad \ldots, \quad A^{n-1}B\mathbf{w},$$

are linearly independent, then the system is controllable.

9. Suppose A is a symmetric matrix with distinct eigenvalues $\{\lambda_j\}$ and eigenvectors $\{\mathbf{x}_{jk}\}$, $k = 1, \ldots, n_j$, n_j is the multiplicity of λ_j, $j = 1, \ldots, P$. If $B = [\mathbf{b}_1, \ldots, \mathbf{b}_m]$ is written in columns, show that $\dot{\mathbf{x}} = A\mathbf{x} + B\mathbf{u}$ is controllable if and only if

$$\operatorname{rank} \begin{bmatrix} \mathbf{b}_1 \cdot \mathbf{x}_{j1} & \cdots & \mathbf{b}_m \cdot \mathbf{x}_{j1} \\ \mathbf{b}_1 \cdot \mathbf{x}_{j2} & \cdots & \mathbf{b}_m \cdot \mathbf{x}_{j2} \\ \vdots & & \vdots \\ \mathbf{b}_1 \cdot \mathbf{x}_{jn_j} & \cdots & \mathbf{b}_m \cdot \mathbf{x}_{jn_j} \end{bmatrix} = n_j, \qquad j = 1, 2, \ldots, P.$$

In particular, for the system to be controllable, the number of controls must exceed the largest multiplicity of the eigenvalues of A.

References

[1] F. Gantmacher, "The Theory of Matrices." Chelsea, New York, 1959.
[2] H. R. Hermes and J. P. LaSalle, "Functional Analysis and Time-Optimal Control." Academic Press, New York, 1969.
[3] D. Russell, Exact boundary value controllability theorems for wave and heat processes in star-complemented regions, *in* "Differential Games and Control Theory" (E. O. Roxin, P. T. Liu, and R. L. Sternberg, eds.). Dekker, New York, 1974.

Chapter VIII State-Constrained Control Problems

1. THE RESTRICTED MAXIMUM PRINCIPLE

In many applications of control theory, some or all of the state variables will be subject to natural constraints, without which the problem may lose its physical meaning. For instance, in the moon-landing problem a natural state constraint would be

$$h(t) \geq 0 \qquad \text{for all} \quad t > 0. \tag{1}$$

In that example the solution was sufficiently simple that we could check and make sure that (1) was not violated during the construction of the optimal solution. In this chapter we shall explore an extension of the maximum principle that will directly cover such state constraints as (1).

We shall now restrict the control problem further. In particular, we suppose the restraint set Ω is set of all $\mathbf{u} \in R^m$ for which

$$q_1(\mathbf{u}) \leq 0, \quad q_2(\mathbf{u}) \leq 0, \quad \ldots, \quad q_s(\mathbf{u}) \leq 0 \tag{2}$$

for given continuously differentiable functions $q_1, \ldots, q_s$. In particular, $\Omega = [-1, 1]$ could be written

$$\{u \in R : u^2 - 1 \leq 0\},$$

that is,

$$q_1(u) = u^2 - 1, \qquad u \in R.$$

We shall place further restrictions on $\{q_1, \ldots, q_s\}$ later.

For the state constraints, we suppose $\mathbf{x}(t)$ is required to lie in a closed region B of the state space R^n of the form

$$B = \{\mathbf{x} : g(\mathbf{x}) \leq 0\} \tag{3}$$

for some given scalar-valued function g, having continuous second partial derivatives, and for which

$$\boldsymbol{\nabla} g = \left(\frac{\partial g}{\partial x_1}, \ldots, \frac{\partial g}{\partial x_n}\right)$$

does not vanish on the boundary of B, that is, on $\{\mathbf{x} : g(\mathbf{x}) = 0\}$. The other elements of the control problem will remain as before.

If we define

$$\rho(\mathbf{x}, \mathbf{u}) = \boldsymbol{\nabla} g(\mathbf{x}) \cdot \mathbf{f}(\mathbf{x}, \mathbf{u}), \tag{4}$$

then once a trajectory $\mathbf{x}(t)$, with control $\mathbf{u}(t)$, hits the boundary of B, $g(\mathbf{x}) = 0$, a necessary and sufficient condition for it to remain there for all later t is that

$$\rho(\mathbf{x}(t), \mathbf{u}(t)) = 0. \tag{5}$$

This asserts just that the velocity of a point moving along the trajectory is tangent to the boundary at time t. Two further assumptions that will be required later are that

$$\boldsymbol{\nabla}_{\mathbf{u}} \rho(\mathbf{x}, \mathbf{u}) \neq \mathbf{0}, \tag{6}$$

and that

$$\boldsymbol{\nabla}_{\mathbf{u}} \rho(\mathbf{x}, \mathbf{u}), \quad \boldsymbol{\nabla}_{\mathbf{u}} q_1(\mathbf{u}), \quad \ldots, \quad \boldsymbol{\nabla}_{\mathbf{u}} q_s(\mathbf{u}) \tag{7}$$

be linearly independent along an optimum trajectory $\mathbf{x}^*(t)$, $\mathbf{u}^*(t)$.

Our strategy for solving state-constrained problems will be to break them into two subproblems. Whenever the optimum trajectory $\mathbf{x}^*(t)$ lies in the interior of B, $g(\mathbf{x}^*(t)) < 0$, the state constraints are superfluous (nonbinding), and we can apply the maximum principle as before. However, as soon as $\mathbf{x}^*(t)$ hits the boundary, $g(\mathbf{x}^*(t)) = 0$, we solve for $\mathbf{x}^*$ using the restricted maximum principle given below. In this way, as long as $\mathbf{x}^*$ consists of a finite number of pieces of these two types, we can hope to construct it.

The essential point of the restricted maximum principle is that once we hit the boundary we have to restrict the control power to prevent the optimum trajectory from leaving B. From our remarks above, this requires that the optimal control $\mathbf{u}^*(t)$ not only take values in Ω, but also that

$$\rho(\mathbf{x}^*(t), \mathbf{u}^*(t)) = 0. \tag{8}$$

So our restricted maximum principle will take into account the added control restraint (8). The Hamiltonian $H(\mathbf{x}, \mathbf{u}, \boldsymbol{\psi})$ is defined as before. A detailed proof of the theorem 1 is given in [4], Chap. VI; see also [2], Chap. VIII, for an alternative treatment.

Theorem 1 (*Restricted Maximum Principle*) Let $\mathbf{x}^*(t)$, $t_0 \leq t \leq t_1$, be an optimal trajectory with control $\mathbf{u}^*(t)$, and suppose that $\mathbf{x}^*(t)$ lies entirely on the boundary of B, $t_0 \leq t \leq t_1$, and (6)–(8) are satisfied. Then there exists a continuous vector function $\boldsymbol{\psi}(t)$ and a piecewise smooth scalar function $\lambda(t)$, $t_0 \leq t \leq t_1$, such that

(i) $$\dot{\mathbf{x}}^*(t) = \mathbf{f}(\mathbf{x}^*(t), \mathbf{u}^*(t)),$$

(ii) $$\dot{\boldsymbol{\psi}}(t) = -\nabla_{\mathbf{x}} H(\mathbf{x}^*(t), \mathbf{u}^*(t), \boldsymbol{\psi}(t)) + \lambda(t)\, \nabla_{\mathbf{x}} \rho(\mathbf{x}^*(t), \mathbf{u}^*(t)),$$

(iii) $$H(\mathbf{x}^*(t), \mathbf{u}^*(t), \boldsymbol{\psi}(t)) = \max_{\mathbf{u}} H(\mathbf{x}^*(t), \mathbf{u}, \boldsymbol{\psi}(t)) = 0,$$

for $t_0 \leq t \leq t_1$, where the maximum is taken subject to $\mathbf{u} \in \Omega$, that is, (2) *and* $\rho(\mathbf{x}^*(t), \mathbf{u}) = 0$. Since this is really just a constrained maximization, we can alternatively write (iii)

(iv) $$\nabla_{\mathbf{u}} H(\mathbf{x}^*(t), \mathbf{u}^*(t), \boldsymbol{\psi}^*(t)) = \lambda(t)\, \nabla_{\mathbf{u}} \rho(\mathbf{x}^*(t), \mathbf{u}^*(t)) + \sum_{r=1}^{s} v_r(t)\, \nabla_{\mathbf{u}} q_r(\mathbf{u}^*(t)), \qquad t_0 \leq t \leq t_1,$$

where $\lambda(t), v_1(t), \ldots, v_s(t)$ are Lagrange multipliers, at least at points t of differentiability of λ. [λ is the same quantity in (ii) and (iv).] Further,

(v) $$\psi_0(t) = \text{const} \leq 0,$$

(vi) $$\boldsymbol{\psi}(t_0) \neq \mathbf{0},$$

and $\boldsymbol{\psi}(t_0)$ is tangent to the boundary of B at $\mathbf{x}^*(t)$.

(vii) Whenever λ is differentiable

$$\dot{\lambda}(t)\, \nabla g(\mathbf{x}^*(t))$$

is directed toward the interior of B or else is zero.

From assumptions (6)–(8), we can discuss (iv) in more detail. At any point of continuity of λ, for given $\mathbf{x}^*(t)$, $\mathbf{u}^*(t)$, and $\boldsymbol{\psi}(t)$, (iv) is a system of m equations in $s + 1$ unknowns $\lambda(t), v_1(t), \ldots, v_s(t)$. Assumptions (6) and (7) guarantee that the matrix of this system has rank $s + 1$, and so we can solve for these unknowns; in particular, we can solve for λ as

$$\lambda(t) = \boldsymbol{\psi}(t) \cdot \mathbf{a}(t), \qquad t_0 \le t \le t_1, \tag{9}$$

where $\mathbf{a}$ is a piecewise smooth function. If (9) is substituted back into (ii), we see that $\boldsymbol{\psi}$ satisfies a linear homogeneous differential equation, and since $\boldsymbol{\psi}(t_0) \neq \mathbf{0}$, we have $\boldsymbol{\psi}(t) \not\equiv \mathbf{0}$ on any interval.

2. JUMP CONDITIONS

So far we know how to compute each piece of the optimum trajectory; all that is lacking is what happens when we change from $g(\mathbf{x}^*(t)) < 0$ to $g(\mathbf{x}^*(t)) = 0$, or vice versa. In particular, what happens to the adjoint variables when we enter the boundary of B, or when we leave it? It can be shown that if t_0 is a time when $\mathbf{x}^*$ enters the boundary, then

$$\boldsymbol{\psi}(t_0 - 0) = \boldsymbol{\psi}(t_0 + 0), \tag{1}$$

that is, $\boldsymbol{\psi}$ is continuous across the jump. On the other hand, on leaving the boundary at some time t_1, say, it can be shown that

$$\boldsymbol{\psi}(t_1 + 0) = \boldsymbol{\psi}(t_1 - 0) - \lambda(t_1)\,\nabla g(\mathbf{x}^*(t_1)), \tag{2}$$

λ as in (ii) and (iv) in Section 1. Conditions (1) and (2) are called the *jump conditions* [4, Chapter 6].

Example 1 Let us consider the application of the previous theorems to a simple "obstacle" problem. Find the planar curve of minimum length joining two given points, and avoiding a closed circular region (see Fig. 1). Without loss of generality we choose one point at the origin. An equivalent control problem would be to minimize the time taken to traverse the given path at constant speed. If u_i represents the component of the velocity in the x_i direction, $i = 1, 2$, then we would have

$$\dot{x}_1 = u_1, \qquad \dot{x}_2 = u_2, \tag{3}$$

with the *control restraint set*

$$\Omega = \{(u_1, u_2) : u_1^2 + u_2^2 = 1\}, \tag{4}$$

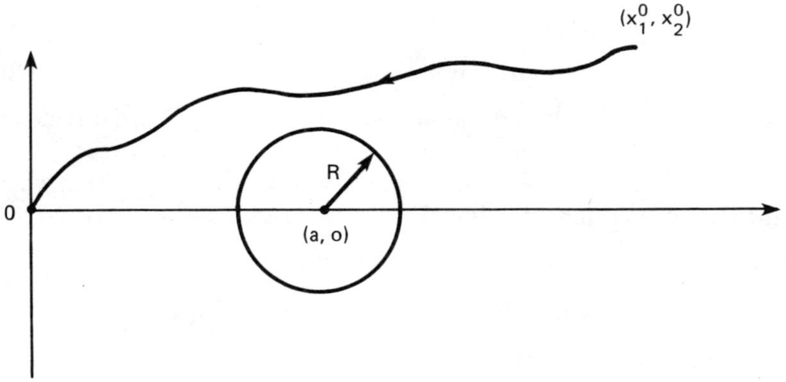

Fig. 1

and *state constraint*

$$g(x_1, x_2) = R^2 - (x_1 - a)^2 - x_2^2 \leq 0. \tag{5}$$

We seek an admissible control, transferring $\mathbf{x}_0$ to $\mathbf{0}$ in minimum time, such that (5) is satisfied for all time, that is, we take as cost

$$C(\mathbf{u}) = \int_0^{t_1} 1\, dt.$$

Consequently, the Hamiltonian is

$$H(\mathbf{x}, \mathbf{u}, \boldsymbol{\psi}) = -1 + \psi_1 u_1 + \psi_2 u_2.$$

First, we analyze the optimum trajectory when the constraint (5) is nonbinding. The adjoint equations are as usual

$$\dot{\psi}_1 = -\frac{\partial H}{\partial x_1} = 0,$$

$$\dot{\psi}_2 = -\frac{\partial H}{\partial x_2} = 0,$$

which can be solved to give

$$\psi_1 = c_1, \qquad \psi_2 = c_2,$$

where c_1, c_2 are constants. The Hamiltonian is maximized when

$$u_i(t) = \frac{c_i}{\sqrt{c_1^2 + c_2^2}}, \qquad i = 1, 2,$$

However, from (iii) of Section 1, $-1 + \sqrt{c_1^2 + c_2^2} = 0$, so

$$u_i(t) = c_i = \psi_i(t), \qquad i = 1, 2, \tag{6}$$

where $\sqrt{c_1^2 + c_2^2} = 1$, and the optimum trajectory inside the region B is just a straight line, as we would expect.

Now we apply the restricted maximum principle in the region where the state constraint is binding. First,

$$\nabla g = (-2(x_1 - a), -2x_2),$$

$$\rho(\mathbf{x}, \mathbf{u}) = \nabla g \cdot \mathbf{f} = -2(x_1 - a)u_1 - 2x_2 u_2,$$

and so $\rho(\mathbf{x}, \mathbf{u}) = 0$ when

$$\frac{u_2}{u_1} = -\frac{x_1 - a}{x_2}.$$

Since $(x_1 - a)^2 + x_2^2 = R^2$ and $u_1^2 + u_2^2 = 1$, this reduces to

$$u_1 = \frac{-x_2}{R}, \qquad u_2 = \frac{x_1 - a}{R}. \tag{7}$$

The adjoint equations now become

$$\dot{\boldsymbol{\psi}} = 0 + \lambda \begin{bmatrix} -2u_1 \\ -2u_2 \end{bmatrix}$$

or (8)

$$\dot{\psi}_1 = -2\lambda u_1,$$
$$\dot{\psi}_2 = -2\lambda u_2.$$

The constrained maximum of the Hamiltonian will be sought by the unconstrained maximization over all $(u_1, u_2) \in R^2$ of the function

$$H - \lambda\rho + v(u_1^2 + u_2^2 - 1),$$

where λ and v are Lagrange multipliers. Taking partial derivatives with respect to u_1, u_2 gives

$$\begin{bmatrix} \psi_1 \\ \psi_2 \end{bmatrix} - \lambda \begin{bmatrix} -2(x_1 - a_1) \\ -2x_2 \end{bmatrix} + v \begin{bmatrix} 2u_1 \\ 2u_2 \end{bmatrix} = 0.$$

Since u_1 and u_2 can be expressed in terms of x_1 and x_2 by (7), if we substitute for x_1 and x_2, we find

$$\psi_1 + 2R\lambda u_2 + 2vu_1 = 0,$$
$$\psi_2 - 2R\lambda u_1 + 2vu_2 = 0.$$

Now eliminating v from these equations gives

$$\psi_1 u_2 - \psi_2 u_1 + 2\lambda R(u_1^2 + u_2^2) = 0,$$

and since $u_1^2 + u_2^2 = 1$, we have

$$\lambda = \frac{\psi_2 u_1 - \psi_1 u_2}{2R}. \tag{9}$$

Further, the second part of (iii) in Section 1 tells us that

$$\psi_2(t)u_2(t) + \psi_1(t)u_1(t) = 1 \tag{10}$$

for the optimum u_1, u_2.

We now have four equations (8)–(10), and the control restraint $u_1^2 + u_2^2 = 1$ for the five unknowns ψ_1, ψ_2, u_1, u_2, λ. Consequently, we can solve for the adjoint variables ψ_1 and ψ_2 and find the optimal solution.

We do this by changing to polar coordinates (Fig. 2). Namely, since $(x_1(t) - a)^2 + x_2^2(t) = R^2$, we let,

$$\begin{aligned} x_1(t) - a &= R\cos\theta(t), \\ x_2(t) &= R\sin\theta(t). \end{aligned} \tag{11}$$

Then

$$\begin{aligned} u_1 &= -\sin\theta, \\ u_2 &= \cos\theta \end{aligned} \tag{12}$$

by (7), and

$$\begin{aligned} \dot{x}_1 &= u_1 = (R\sin\theta)\dot{\theta}, \\ \dot{x}_2 &= u_2 = -(R\cos\theta)\dot{\theta}, \end{aligned}$$

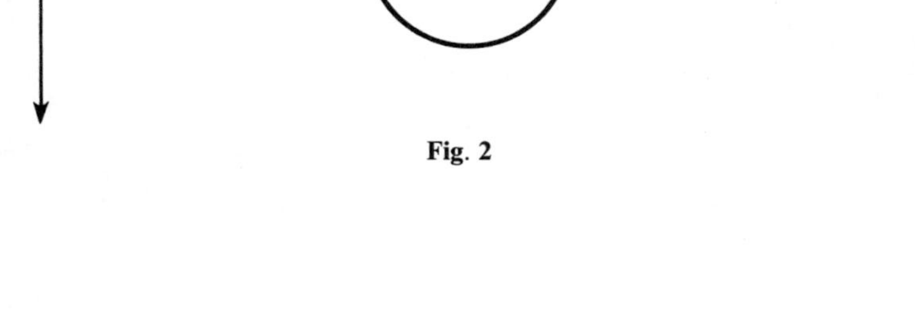

Fig. 2

so squaring and adding gives

$$\dot{\theta} = 1/R.$$

By the chain rule applied to (8), we can solve for ψ_1 in terms of θ, that is,

$$\frac{d\psi_1}{d\theta} = \frac{d\psi_1}{dt}\frac{dt}{d\theta} = R\frac{d\psi_1}{dt} = \psi_1 \tan\theta + \tan\theta \sin\theta.$$

Integrating gives

$$\psi_1(\theta) = K\cos\theta - \sin\theta + \theta\cos\theta. \tag{13}$$

Similarly,

$$\psi_2(\theta) = K\sin\theta + \cos\theta + \theta\sin\theta. \tag{14}$$

To describe the behavior of $\boldsymbol{\psi}$ as we enter and leave the circle of radius R, we now apply the jump conditions. At an entering point $t = t_0$, say, $\theta(t_0) = \theta_0$, we have

$$\begin{aligned} K\cos\theta_0 - \sin\theta_0 + \theta_0\cos\theta_0 &= -\cos\varphi_0, \\ K\sin\theta_0 + \cos\theta_0 + \theta_0\sin\theta_0 &= \sin\varphi_0, \end{aligned} \tag{15}$$

where, by (6), we have let $\psi_1(t) = -\cos\varphi_0$, $0 \le t \le t_0$, and $\psi_2(t) = \sin\varphi_0$, $0 \le t \le t_0$, with φ_0 as shown in Fig. 3. Solving (15) gives

$$K = -\theta_0 \tag{16}$$

and

$$\theta_0 + \varphi_0 = \pi/2. \tag{17}$$

In other words, *the portion of an optimal trajectory that precedes an entering state is tangent to the boundary*. Employing the jump condition

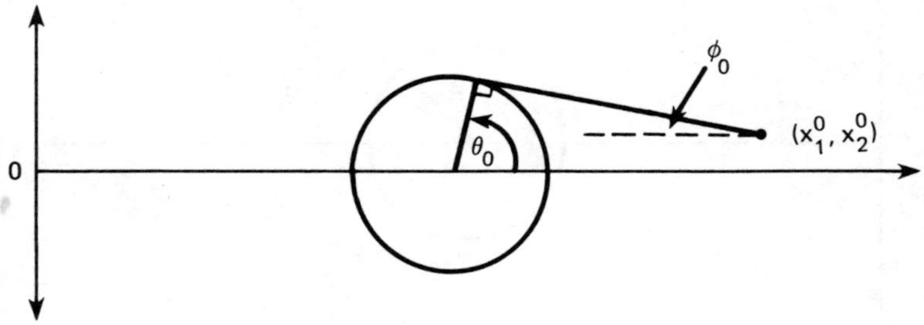

Fig. 3

for a leaving point $t = t_1$, $\theta(t_1) = \theta_1$, we have

$$\begin{aligned} \psi_1(\theta_1 - 0) &= -\sin\theta_1 \\ \psi_2(\theta_1 - 0) &= \cos\theta_1 \end{aligned} \qquad \text{[using (16) in (13) and (14)]},$$

and

$$\begin{aligned} \lambda(\theta_1) &= \frac{\psi_2(\theta_1)u_1(\theta_1) - \psi_1(\theta_1)u_2(\theta_1)}{2R} \\ &= \frac{(\cos\theta_1)(-\sin\theta_1) - (-\sin\theta_1)(\cos\theta_1)}{2R} \\ &= 0. \end{aligned}$$

So the jump (2) in the adjoint variables is 0 at $\theta = \theta_1$, and

$$\begin{aligned} \psi_1(\theta_1 + 0) &= -\sin\theta_1, \\ \psi_2(\theta_1 + 0) &= \cos\theta_1. \end{aligned} \tag{18}$$

Since the trajectory leaving the boundary is again a straight line, we have as before

$$\begin{aligned} \psi_1(\theta_1 + 0) &= u_1(\theta_1 + 0) = -\cos\varphi_1, \\ \psi_2(\theta_1 + 0) &= u_2(\theta_1 + 0) = -\sin\varphi_1, \end{aligned} \tag{19}$$

where φ_1 is the angle shown in Fig. 4. Combining (18) and (19) gives

$$\begin{aligned} \cos\varphi_1 &= \sin\theta_1, \\ \sin\varphi_1 &= -\cos\theta_1, \end{aligned}$$

so, again, *the optimum trajectory leaves tangential to the boundary.*

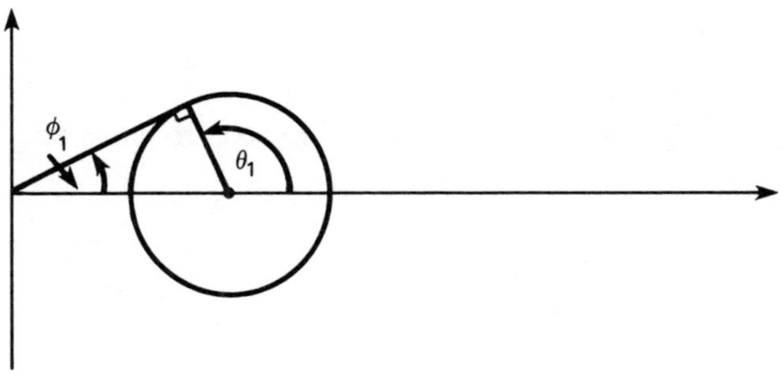

Fig. 4

Summing up, the optimum trajectory consists of straight lines; if it touches the boundary, it follows the boundary and leaves and enters tangentially.

3. THE CONTINUOUS WHEAT TRADING MODEL WITHOUT SHORTSELLING†

We now return to the control problem in Example 4.2.1, namely,

$$\text{maximize } C = x_1(T) + p(T)x_2(T) \tag{1}$$

subject to

$$\dot{x}_1(t) = -p(t)u(t) - \alpha x_2(t), \qquad x_1(0) \text{ given}, \tag{2}$$

$$\dot{x}_2(t) = u(t), \qquad x_2(0) \text{ given}. \tag{3}$$

We recall that $x_1(t)$ and $x_2(t)$ represented the firm's stock of cash and wheat, respectively, α is the unit holding cost of wheat storage, and p is the price of wheat over the planning horizon $[0, T]$, assumed continuous. The control function u represents the rate of sale or procurement of wheat, and we suppose without loss of generality that

$$-1 \leq u(t) \leq 1, \qquad t \geq 0. \tag{4}$$

Since we forbid shortselling, we have the state constraint

$$x_2(t) \geq 0, \qquad t \geq 0, \tag{5}$$

and so the feasible region is $B = \{(x_1, x_2) : x_2 \geq 0\}$. For this problem,

$$p(\mathbf{x}, u) = -u. \tag{6}$$

Now let $\mathbf{x}^*(t) = (x_1^*(t), x_2^*(t))$ be an optimal trajectory with control $u^*(t)$, and suppose $\mathbf{x}^*(t)$ lies entirely on the boundary of B, for $t_k \leq t \leq t_{k+1}$. Trivially, from (6) the optimal control must be

$$u^*(t) = 0, \qquad t_k \leq t \leq t_{k+1}. \tag{7}$$

Substituting (7) into (2) and (3) gives

$$x_1^*(t) = x_1^*(t_k), \qquad x_2^*(t) = 0, \qquad t_k \leq t \leq t_{k+1}.$$

† See Noström [3].

Since the control constraints are now no longer binding, their respective multipliers in (1.12) are zero, and so we have

$$\lambda(t) = \varphi_1(t)p(t) - \varphi_2(t), \qquad t_k \leq t \leq t_{k+1}.$$

Finally, the adjoint equations for the constrained problem are unchanged from before,

$$\dot{\varphi}_1(t) = 0, \qquad \dot{\varphi}_2(t) = \alpha\varphi_1(t), \qquad t_k \leq t \leq t_{k+1},$$

with the transversality conditions $\varphi_1(T) = -1$, $\varphi_2(T) = -p(T)$. From the jump conditions both φ_1 and φ_2 are continuous at $t = t_k$. At $t = t_{k+1}$,

$$\varphi_1(t_{k+1}^+) = \varphi_1(t_{k+1}^-),$$

so by the transversality conditions,

$$\varphi_1(t) \equiv -1, \qquad 0 \leq t \leq T.$$

Next,

$$\varphi_2(t_{k+1}^+) = \varphi_2(t_{k+1}^-) + \lambda(t_{k+1}^-) = -p(t_{k+1}),$$

so the adjoint variable $-\varphi_2$ may have a jump at $t = t_{k+1}$ and is, after the jump, equal to the price of wheat there, $p(t_{k+1})$. It also follows from the continuity of the Hamiltonian ((iii), Section 1) that

$$\varphi_2(t_k) = \varphi_2(t_k^-) = -p(t_k^-) = -p(t_k).$$

To give a complete solution of this problem, we now suppose that the optimal trajectory consists of a finite number of constrained and unconstrained arcs. From the solution obtained below it is not hard to see this will be the case if p consists of a finite number of concave or convex segments. This case would cover most price functions of interest.

In any time period $[t_{k-1}, t_k]$ where the trajectory leaves the boundary at time t_{k-1} and enters again at time t_k, the state constraint will be nonbinding and the solution will be as before (Example 4.2.1), with $p(T)$ replaced by $p(t_k)$. It only remains to find the points $t_1, t_2, \ldots, t_N$ where the trajectory enters or leaves the boundary of B. To find these points, we use two additional facts. First, since $x_2^*(t_{k-1}) = x_2^*(t_k) = 0$, the amount of wheat bought in an interval $[t_{k-1}, t_k]$ must exactly equal the amount sold. Second, by (vii) of Section 1 on boundary arcs we must have

$$\dot{p}(t) < \alpha, \tag{8}$$

at points where p is differentiable, and so the points $t_1, \ldots, t_N$ must lie in intervals where (8) holds.

To solve the problem given a function p is now largely geometric. We solve the example in Chapter IV, Section 2, in Fig. 5. The final arc of φ_2 is determined by the transversality condition, $-\varphi_2(T) = p(T)$, and the fact that its slope is α. If we draw this, we find the point t_4. The two requirements above and the fact that the slope of φ_2 is α lead to t_2 and t_3 and to the switch from buying to selling at $t = 3$. We continue along in this fashion until arriving at $t = 0$. The optimal control is

$$u^*(t) = \begin{cases} -1 \quad \text{(sell)}, & 0 \le t < 1 \\ 0 \quad \text{(constrained)}, & 1 \le t \le 1.8 \\ 1 \quad \text{(buy)}, & 1.8 < t < 3 \\ -1 \quad \text{(sell)}, & 3 < t < 4.2 \\ 0 \quad \text{(constrained)}, & 4.2 \le t \le 4.6 \\ 1 \quad \text{(buy)}, & 4.6 < t < 7. \end{cases}$$

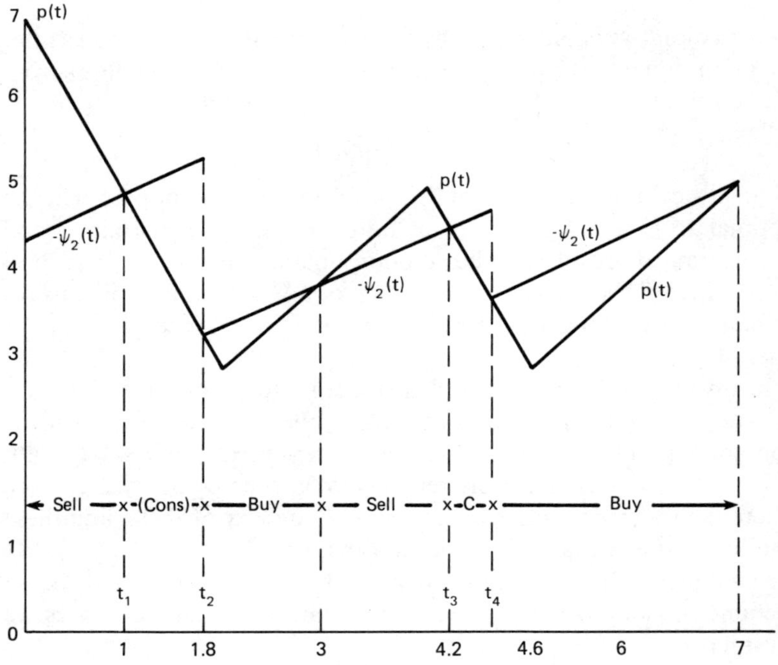

Fig. 5

4. SOME MODELS IN PRODUCTION AND INVENTORY CONTROL

We consider an inventory of one or more items being controlled over a fixed time period $[0, T]$. At any time t, the rate of change of inventory level will be the difference between orders or production quantities received and the demand fulfilled. In other words

$$\dot{x}(t) = u(t) - d(t), \qquad x(0) = x_0 > 0, \tag{1}$$

where $x(t)$ is the inventory at time t, $u(t)$ the production rate, and $d(t)$ the demand rate, which is assumed positive and given. As the cost of holding inventory $x(t)$ and producing at a rate $u(t)$ we take

$$C(u) = \int_0^T (cu(t) + hx(t))\, dt \tag{2}$$

where c and h are the unit ordering and holding costs per unit time. We assume that they are positive and constant, so that the unit costs depend neither on time nor on the quantity of goods ordered or held. *We seek a control (production rate) u that minimizes this cost.*

We also assume state and control constraints of the form

$$x(t) \geq 0, \tag{3}$$

$$u(t) \geq 0, \tag{4}$$

so that backlogging of inventory and disposal of production will not be allowed.

To begin with, let us apply the necessary conditions in the case in which the state constraint (3) is nonbinding,

$$H(x, u, \varphi) = -cu - hx + \varphi(u - d), \tag{5}$$

and the adjoint equation is

$$\dot{\varphi} = -\frac{\partial H}{\partial x} = h. \tag{6}$$

The maximization of the Hamiltonian is equivalent to,

$$\max_{u \geq 0} \{u(\varphi - c)\}. \tag{7}$$

From (7), we see that if $\varphi - c > 0$, the maximum $= +\infty$, so no optimal control would exist. If $\varphi(t) - c \leq 0$, the maximum occurs with $u^*(t) = 0$. However, we must check that if $x^*(t) > 0$ and $u^*(t) = 0$, then $\varphi(t) - c$

is indeed less than or equal to zero. To do this, we recall that the maximum principle states that along an optimal trajectory

$$H(x^*(t), u^*(t), \varphi(t)) = 0$$

so

$$-hx^*(t) - \varphi(t)d(t) = 0. \tag{8}$$

Now if $u^*(t) \equiv 0$, then $\dot{x}^*(t) = -d(t) \leq 0$, so $x^*(t)$ is decreasing from its positive initial value x_0 as t increases. For the moment, let us assume $x^*(t)$ reaches zero for $t < T$. Namely, let $t_0 < T$ be the first time that

$$x_0 = \int_0^{t_0} d(t)\,dt. \tag{9}$$

Then $x(t_0) = 0$, and so by (8), $\varphi(t_0) = 0$. Consequently, from (6)

$$\varphi(t) = h(t - t_0), \qquad 0 \leq t \leq t_0, \tag{10}$$

and then clearly

$$\varphi(t) - c \leq 0, \qquad 0 \leq t \leq t_0.$$

Under the assumption (9), the necessary conditions are satisfied by the following strategy: *If we begin with positive inventory, we must produce nothing until the inventory level reaches zero.* At this point, however, the state constraint will be violated if we continue this strategy; we must now apply the restricted *maximum principle.*

Note that the state constraint is

$$g(x(t)) \equiv -x(t) \leq 0,$$

so that the function $p(x, u) = \nabla g \cdot \mathbf{f}$ becomes just

$$p(x, u) = -(u - d). \tag{11}$$

Consequently, $\partial p/\partial x = 0$, and the adjoint equation is again

$$\dot{\varphi}(t) = h, \qquad t_0 \leq t. \tag{12}$$

Since the jump condition guarantees continuity of φ across $t = t_0$, and φ has slope h on both sides, we must still have

$$\varphi(t) = h(t - t_0), \qquad t \geq t_0, \tag{13}$$

at least until the optimum trajectory leaves the x axis.

The maximum condition now becomes

$$\text{maximize } H(x^*(t), u, \varphi(t)) \tag{14}$$

subject to $p(x^*(t), u) = 0$ and $u \geq 0$. The first condition implies

$$u^*(t) = d(t), \qquad t > t_0, \tag{15}$$

and since we assumed $d(t) > 0$, the second is automatically satisfied. Hence the new necessary conditions imply that on the x axis we should use the optimal control (15). Note that if we do this,

$$\dot{x}^*(t) = d(t) - d(t) = 0, \qquad x^*(t_0) = 0$$

so that

$$x^*(t) = 0, \qquad t_0 \leq t \leq T,$$

and (15) must be the optimal control over the remaining time interval $[t_0, T]$.

The structure of the optimal policy is therefore to produce nothing until the initial inventory is depleted and then to produce to demand until the end of the process.

$$x^*(t) = \begin{cases} x_0 - \int_0^t d(t)\,dt & 0 \leq t \leq t_0 \\ 0 & t_0 \leq t \leq T, \end{cases} \tag{16}$$

and the multiplier λ is

$$\lambda(t) = c + h(t - t_0), \qquad t_0 \leq t < T. \tag{17}$$

Finally, we remark that if (9) is never satisfied on $[0, T]$, then $x_*(t) > 0$ for all $0 \leq t \leq T$, so we really have a free-end-point fixed-time (unconstrained) control problem. Then the transversality conditions imply $\varphi(T) = 0$, and so $\varphi(t) - c \leq 0, 0 \leq t \leq T$. In other words, our optimal policy is unchanged.

Note that the optimum policy over the time interval $[0, t_0]$ is independent of the demand, in the sense that we need not know the demand function a priori. We simply produce nothing until our inventory runs out; then over the time interval $[t_0, T]$ we must know the demand if we are to operate optimally. In other words, *the optimum policy up to zero inventory is unaffected by the future demand pattern.* This concept is known as a "*planning horizon.*" We remark here that it has been shown that planning horizons exist for more-complicated cost functionals for the inventory problem [1].

Inventory Control with Instantaneous Ordering. We now consider the inventory problem in which orders are placed and received at distinct times during the process, not continuously as before.

Suppose $x(t)$ is again our inventory, which we replenish by ordering amount u_i at time t_1. We assume they ordering is instantaneous. Assume that we place N such orders $u_0, u_1, \ldots, u_{N-1}$ in some fixed time interval $[0, T]$. We assume that $x(0) = 0$, u_0 occurs at $t_0 = 0$, and $t_{N-1} = T$. We let the demand rate be $d(t)$, and denote the cumulative demand up to time t by

$$D(t) = \int_0^t d(t)\,dt, \quad t \geq 0. \tag{18}$$

The costs we face will be the cost of holding level $x(t)$ over $0 \leq t \leq T$. We take this of the form

$$\int_0^T h(t)x(t)\,dt$$

for some given function $h \geq 0$. The ordering costs will be assumed independent of the order size, and given by some constant K; the N orders then cost KN. Our total cost is then

$$C = KN + \int_0^T h(t)x(t)\,dt. \tag{19}$$

The state equations could be written

$$\dot{x}(t) = u_0\,\delta(t) + \sum_{i=1}^{N-1} u_i\,\delta_i(t - t_i) - d(t), \tag{20}$$

$$x(0) = 0, \tag{21}$$

$$x(t) \geq 0, \tag{22}$$

where $\delta(t)$ is the Dirac δ function at 0. The control problem is to choose N, $\{t_i\}$, and $\{u_i\}$ so as to minimize (19).

We shall first rearrange the state equations (20) to make them more amenable. Denote $\tau_i = t_i - t_{i-1}$, $i = 1, 2, \ldots, N$; then

$$\tau_i \geq 0, \qquad \sum_{i=1}^{N} \tau_i = T. \tag{23}$$

If we let $x_i(s)$ be the inventory level in the ith time period $t_{i-1} \leq s \leq t_i$, the state equations can be rewritten

$$\begin{aligned} \frac{dx_i}{ds} &= -\tau_i d(t_{i-1} + s\tau_i), \qquad 0 < s < 1, \\ x_i(0) &= x_{i-1}(1) + u_{i-1}, \qquad i = 1, 2, \ldots, N, \end{aligned} \tag{24}$$

with $x_0(0) = x_0 = 0$. This gives N continuous differential equations on $[0, 1]$ with jumps at the end points.

The solution of (24) is then just the difference of the total number of orders received and the cumulative demand, that is,

$$x_i(s) = \sum_{j=0}^{i-1} u_j - D(t_{i-1} + s\tau_i), \qquad 0 \leq s \leq 1, \tag{25}$$

and by (22),

$$x_i(s) \geq 0, \qquad i = 1, 2, \ldots, N. \tag{26}$$

If we insert (25) into the cost (19), and break the integral over $[0, T]$ into the sum of integrals over $[t_{i-1}, t_i]$, $i = 1, 2, \ldots, N$, then

$$\begin{aligned} C &= KN + \int_0^{t_1} hx\,dt + \int_{t_1}^{t_2} hx\,dt + \cdots + \int_{t_{N-1}}^{T} hx\,dt \\ &= KN + \sum_{i=1}^{N} \int_0^1 h(t_{i-1} + s\tau_i)x_i(s)\,ds \\ &= KN + \sum_{i=1}^{N} \int_0^1 h(t_{i-1} + s\tau_i)\left[\sum_{j=0}^{i-1} u_j - D(t_{i-1} + s\tau_i)\right] ds. \end{aligned} \tag{27}$$

Now by (25) and (26)

$$\sum_{j=0}^{i-1} u_j \geq D(t_{i-1} + s\tau_i), \qquad i = 1, 2, \ldots, N, \quad 0 \leq s \leq 1,$$

so we minimize C by making the difference

$$\sum_{j=0}^{i-1} u_j - D(t_{i-1} + s\tau_i) \tag{28}$$

as small as possible over $0 \leq s \leq 1$. Since D is the indefinite integral of a positive function,

$$D(t_i) \geq D(t_{i-1} + s\tau_i), \qquad 0 \leq s \leq 1,$$

the smallest we can make (28) over $0 \leq s \leq 1$ would be by choosing the optimum $\{u_j\}$ and $\{u_j^*\}$ by

$$\sum_{j=0}^{i-1} u_j^* = D(t_i), \qquad i = 1, 2, \ldots, N. \tag{29}$$

In other words,

$$\begin{aligned} u_0^* &= D(t_1), \\ u_i^* &= D(t_{i+1}) - D(t_i), \qquad i = 1, 2, \ldots, N-1, \end{aligned} \tag{30}$$

and we should order just enough at any order time to last until the next order is received.

Inserting this into (27) gives the cost

$$C = KN + \left(\sum_{i=1}^{N} D(t_i) \int_{t_{i-1}}^{t_i} h(t)\, dt \right) - \int_0^T h(t)D(t)\, dt \tag{31}$$

as a function $C = C(t_i, \ldots, t_N)$, which we can minimize subject to the constraints

$$0 = t_0 < t_1 < t_2 < \cdots < t_{N-1} < t_N = T.$$

It can be shown that the optimal order times t_i^* are then solutions of

$$d(t_i^*) \int_{t_{i-1}^*}^{t_i^*} h(t)\, dt = h(t_i^*)[D(t_{i+1}^*) - D(t_i^*)], \quad i = 1, 2, \ldots, N-1. \tag{32}$$

Note that for a given N, (32) is an $(N-1)$-dimensional system of nonlinear equations. In general we cannot determine N a priori, so we guess a value of N, solve (32), and compute the minimum cost (31). Then we repeat the method with larger N until the cost no longer reduces.

In the case $h(t) = h = \text{const}$, $d(t) = d = \text{const}$, (32) then becomes

$$dh(t_i^* - t_{i-1}^*) = h[dt_{i+1}^* - dt_i^*]$$

or

$$t_i^* - t_{i-1}^* = t_{i+1}^* - t_i^*, \qquad i = 1, 2, \ldots, N-1.$$

In other words, the order intervals are of constant length, say τ.

The cost then becomes

$$KN + \sum_{i=1}^{N} (h\, d\tau^2) i - \frac{h\, dT^2}{2} = KN - \frac{h\, dT^2}{2} + h\, d\tau^2 \frac{N(N+1)}{2}.$$

Then substituting $N = T/\tau$ and minimizing over τ gives

$$\min\left(\frac{KT}{\tau} + \frac{hd}{2} T(T+\tau) \right),$$

which occurs for

$$-\frac{KT}{\tau^2} + \frac{h\, dT}{2} = 0$$

or

$$\tau^2 = \frac{2K}{hd}, \qquad \tau = \sqrt{\frac{2K}{hd}}.$$

From this value we can calculate N, $\{u_i^*\}$, and C from (30 and (19).

Problems

1. Consider the minimum time control to the origin for the system

$$\ddot{x} = u,$$

with initial conditions,

$$x(0) = x_0, \qquad \dot{x}(0) = 0$$

and the state constraint,

$$|\dot{x}(t)| \leq 1, \qquad t \geq 0.$$

2. Resolve Example 2.1 in the case of the curve merely terminating on the x_2 axis instead of at the origin.

3. Find the control that transfers (0, 0) to the axis $x_1 = a > 0$ in minimum time, for the system

$$\dot{x}_1 = \sqrt{x_2}u_1, \qquad \dot{x}_2 = \sqrt{x_2}u_2, \qquad u_1^2 + u_2^2 = 1,$$

with the state constraint

$$x_2 \leq rx_1 + s, \qquad r, s > 0.$$

References

[1] A. Bensoussan, E. Hurst, and B. Naslund, "Management Applications of Modern Control Theory. North-Holland, Publ., New York, 1974.

[2] M. R. Hestenes, Multiplier and gradient methods, *J. Optim. Theory Appl.* **4**, 303–320 (1969).

[3] C. Norström, The continuous wheat trading model reconsidered. An application of mathematical control theory with a state constraint. Working Paper, Graduate School of Industrial Administration, Carnegie-Mellon University, 1978.

[4] L. Pontryagin, V. Boltyanskii, R. Gramkrelidze, and E. Mischenko, "The Mathematical Theory of Optimal Processes." Wiley (Interscience), New York, 1962.

Chapter IX Optimal Control of Systems Governed by Partial Differential Equations

1. SOME EXAMPLES OF ELLIPTIC CONTROL PROBLEMS

To begin with, consider the following simple elliptic control problem. We are given a region $\Omega \subset R^N$ with sufficiently regular boundary Γ (see Fig. 1).

Suppose the set of admissible controls U, a closed, bounded, convex set in $L^2(\Omega)$, is given, and for each $u \in U$ let $y(u)$ be the solution of the Dirichlet problem [2]

$$\Delta y(u) = f + u \qquad \text{in} \quad \Omega, \tag{1}$$

$$y(u) = 0 \qquad \text{on} \quad \Gamma, \tag{2}$$

where f is a given fixed function in $L^2(\Omega)$. We can ask to find the control $u \in U$ that minimizes the cost

$$C(u) = \int_\Omega |y(u) - y_{\mathrm{d}}|^2 \, d\mathbf{x}. \tag{3}$$

In this case, the cost $C: L^2(\Omega) \to R$, and we ask for $\min_{u \in U} C(u)$.

If we regard $y(u)$ as the steady-state temperature on Ω corresponding to u, then the control problem has the physical interpretation of finding a control u such that the steady-state temperature on Ω is as close as possible (in the mean-square sense) to some desired temperature distribution y_{d}.

Alternatively, we could regard $y(u)$ as the deformation induced in the body Ω by a loading u. Then the control problem would be to find a u that achieves a deformation as close as possible (again in the mean-square sense) to y_{d}.

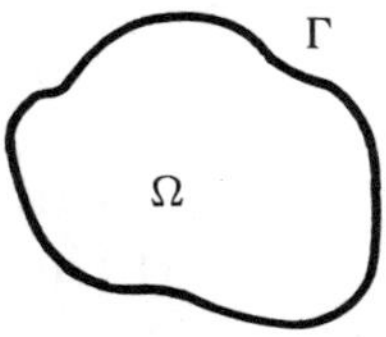

Fig. 1

Necessary conditions for the analogous problem for $U \subset R^n$ can be easily obtained. Namely, suppose that we have a differentiable function $C: U \subset R^n \to R$ and that U is a closed, bounded, convex set. Then the minimum of C exists on U, at say $\mathbf{u}^*$, and since U is convex,

$$C(\mathbf{u}^*) \leq C((1-\theta)\mathbf{u}^* + \theta\mathbf{v}) \qquad \text{for all} \quad \mathbf{v} \in U, \quad 0 \leq \theta \leq 1,$$

that is,

$$C(\mathbf{u}^* + \theta(\mathbf{v} - \mathbf{u}^*)) - C(\mathbf{u}^*) \geq 0, \qquad 0 \leq \theta \leq 1,$$

and so

$$\frac{1}{\theta}[C(\mathbf{u}^* + \theta(\mathbf{v} - \mathbf{u}^*)) - C(\mathbf{u}^*)] \geq 0. \tag{4}$$

As C is differentiable we can let $\theta \to 0$ in (4), and we arrive at the necessary condition for optimality

$$DC(\mathbf{u}^*)(\mathbf{v} - \mathbf{u}^*) \geq 0 \qquad \text{for all} \quad \mathbf{v} \in U.$$

In other words, the directional derivative of C at $\mathbf{u}^*$ in the direction $\mathbf{v} - \mathbf{u}^*$ must be nonnegative for all $\mathbf{v} \in U$.

Of course, in the case $U = R^n$, this condition reduces to the familiar optimality criterion

$$\nabla C(\mathbf{u}^*) = 0.$$

We can certainly try this argument on our cost functional (3); now of course $C: U \subset L^2(\Omega) \to R$. The directional derivative as defined by (4) would just be

$$\lim_{\theta \to 0} \frac{1}{\theta}\left[\int_\Omega [(y(u^* + \theta(v - u^*)) - y_d)^2 - (y(u^*) - y_d)^2]\,d\mathbf{x}\right]$$

$$= \lim_{\theta \to 0} \frac{1}{\theta}\left[\int_\Omega [(y(u^*) + \theta y(v) - \theta y(u^*) - y_d)^2 - (y(u^*) - y_d)^2]\,d\mathbf{x}\right]$$

$$= \lim_{\theta \to 0} \frac{1}{\theta}\left[\int_\Omega [(y(u^*) - y_d)^2 + 2(\theta y(v) - \theta y(u^*))(y(u^*) - y_d) + \theta^2(y(v) - y(u^*))^2 - (y(u^*) - y_d)^2]\,d\mathbf{x}\right].$$

Now canceling and letting $\theta \to 0$, we arrive at

$$\lim_{\theta\to 0}\frac{1}{\theta}\left[\int_\Omega [(y(u^* + \theta(v - u^*)) - y_d)^2 - (y(u^*) - y_d)^2]\,d\mathbf{x}\right]$$
$$= 2\int_\Omega (y(u^*) - y_d)(y(v) - y(u^*))\,d\mathbf{x}.$$

This, then, is the value of the directional derivative of C at u^*, in the direction $v - u^*$, except that now v and u^* are elements of the Hilbert space $L^2(\Omega)$. If u^* is an optimal control, then we must have, as before,

$$\int_\Omega (y(u^*) - y_d)(y(v) - y(u^*))\,d\mathbf{x} \geq 0 \qquad \text{for all} \quad v \in U. \tag{5}$$

We can rearrange this necessary condition to achieve some simplification. Since $y(u) - y_d \in L^2(\Omega)$, we can find a $p(u) \in H^2(\Omega)$ satisfying

$$\begin{aligned} \Delta p(u) &= y(u) - y_d && \text{in} \quad \Omega, \\ p(u) &= 0 && \text{on} \quad \Gamma, \end{aligned}$$

for any $u \in U$. Then in (5) we have by Green's identity

$$\int_\Omega (\Delta p(u^*))(y(v) - y(u^*))\,d\mathbf{x} = \int_\Omega p(u^*)\,\Delta(y(v) - y(u^*))\,d\mathbf{x},$$

since $p(u^*) = y(u^*) = y(v) = 0$ on Γ, and so (5) becomes

$$\int_\Omega p(u^*)(v - u^*)\,d\mathbf{x} \geq 0 \qquad \text{for all} \quad v \in U.$$

Now we can restate the necessary condition (5) as

$$\begin{aligned} \Delta y(u^*) &= f + u^* && \text{in} \quad \Omega, \\ y(u^*) &= 0 && \text{on} \quad \Gamma, \\ \Delta p(u^*) &= y(u^*) - y_d && \text{in} \quad \Omega, \\ p(u^*) &= 0 && \text{on} \quad \Gamma, \end{aligned} \tag{6}$$

and

$$\int_\Omega p(u^*)(v - u^*)\,d\mathbf{x} \geq 0 \qquad \text{for all} \quad v \in U. \tag{7}$$

In the case in which the set of admissible controls is given by pointwise constraints, (7) can be shown to be equivalent to [5, Theorem II, 2.1]

$$p(u^*)(\mathbf{x})(v(\mathbf{x}) - u^*(\mathbf{x})) \geq 0 \qquad \text{a.e.} \quad \mathbf{x} \in \Omega \quad \text{for all} \quad v \in U. \tag{8}$$

Example 1 If $U = \{u : |u(\mathbf{x})| \leq 1, \mathbf{x} \in \Omega\}$, then (7) becomes

$$p(u^*)(\mathbf{x})u^*(\mathbf{x}) \geq p(u^*)(\mathbf{x})v \qquad \text{for all} \quad |v| \leq 1,$$

that is,

$$u^*(\mathbf{x}) = \operatorname{sgn}(p(u^*)(\mathbf{x})) \qquad \text{a.e.} \quad \mathbf{x} \in \Omega. \tag{8'}$$

Unfortunately, we cannot expect $p(u^*)$ to be nonzero almost everywhere, and so u^* in (8′) may not be bang-bang everywhere in Ω.

Of course $p(u^*) \not\equiv 0$ unless $y(u^*) = y_d$, and the problem is trivial. Note, however, that if $p(u^*) = 0$ on $E \subset \Omega$, E nonnull, then

$$\Delta p(u^*) = 0 \qquad \text{a.e. on} \quad E,$$

that is,

$$y(u^*) - y_d = 0 \qquad \text{a.e. on} \quad E$$

and so if y_d is sufficiently regular, i.e., $y_d \in H^2(\Omega)$, then

$$\Delta y(u^*) - \Delta y_d = 0 \qquad \text{a.e.} \qquad \text{and} \qquad f + u^* = \Delta y_d \qquad \text{on} \quad E$$

or

$$u^*(\mathbf{x}) = \Delta y_d(\mathbf{x}) - f(\mathbf{x}) \qquad \text{a.e.} \quad \mathbf{x} \in E. \tag{9}$$

In the case in which $|\Delta y_d(\mathbf{x}) - f(\mathbf{x})| > 1$, (9) is impossible, since $|u^*(\mathbf{x})| \leq 1$. Similarly, if $|\Delta y_d(\mathbf{x}) - f(\mathbf{x})| = 1$, (9) would contradict the definition of the set E (where u^* was not bang-bang). Hence we must have $|u^*(\mathbf{x})| = 1$ for almost all $\mathbf{x} \in \{\mathbf{t} \in \Omega : |\Delta y_d(\mathbf{t}) - f(\mathbf{t})| \geq 1\}$. This gives the following immediate Proposition.

Proposition 1 If $|\Delta y_d(\mathbf{x}) - f(\mathbf{x})| \geq 1$ for almost all $\mathbf{x} \in \Omega$, then u^* is bang-bang.

2. NECESSARY AND SUFFICIENT CONDITIONS FOR OPTIMALITY

In this section we state the control problem somewhat more abstractly in the hope of covering a wider range of applications.

First, let H be a separable Hilbert space with inner product (,) and suppose we are given a continuous, symmetric, positive semidefinite bilinear form $a(u, v)$ on $H \times H$, that is,

bilinear

$$\begin{aligned} a(u_1 + u_2, v) &= a(u_1, v) + a(u_2, v), && u_1, u_2, v \in H, \\ a(u, v_1 + v_2) &= a(u, v_1) + a(a, v_2), && u, v_1, v_2 \in H, \\ a(\alpha u, v) &= \alpha a(u, v) && \alpha \in R, \quad u, v \in H \end{aligned}$$

continuous

$$|a(u, v)| \leq K\|u\|\,\|v\| \qquad \text{for some constant} \quad K,$$

symmetric

$$a(u, v) = a(v, u), \qquad v, u \in H,$$

positive definite

$$a(u, u) > 0 \qquad \text{for all} \quad u \in H, u \neq 0.$$

Suppose we are also given a continuous linear functional $u \to L(u)$, $u \in H$, and a closed convex set $U \subset H$, which we shall consider as the set of admissible controls. Under these assumptions, we take the cost in the control problem to be

$$C(u) = a(u, u) - 2L(u) \qquad \text{for} \quad u \in U. \tag{1}$$

We shall show in Section 3 that this form of the cost functional includes the problem example in Section 1, as well as many other problems of interest.

The following theorem gives some conditions on the form a and the control restraint set U for an optimal control to exist.

Theorem 1 If a is coercive on H, that is,

$$a(u, u) \geq K\|u\|^2 \qquad \text{for all} \quad u \in H, \quad K > 0,$$

then there exists a unique $u^* \in U$ minimizing C over U. If U is bounded, a unique minimizing element always exists (even if a is not coercive).

Proof Suppose a is coercive on U, and let $\{u_n\}$ be a minimizing sequence, that is,

$$C(u_n) \downarrow \inf_{u \in U} C(u).$$

Then

$$\begin{aligned} |C(u_1)| \geq |C(u_n)| &= |a(u_n, u_n) - 2L(u_n)| \\ &\geq |a(u_n, u_n)| - 2|L(u_n)| \\ &\geq K\|u_n\|^2 - 2K_1\|u_n\|, \qquad n = 1, 2, 3, \ldots, \end{aligned}$$

by coercivity and boundedness of L. Solving this quadratic inequality, we must have

$$\|u_n\| \leq K_2,$$

that is, the minimizing sequence $\{u_n\}$ is bounded. Since H is a Hilbert space, the set $\{u_n\}$ is relatively weakly sequentially compact. Consequently, we can choose a weakly convergent subsequence $u_{n_k} \to u^*$, weakly in H, $u^* \in H$. U is closed and convex; hence it is a weakly closed subset of H, and so $u^* \in U$. However, the function $u \to C(u)$ is convex and norm continuous; hence it is weakly lower semicontinuous, and so

$$C(u^*) \leq \liminf_{k\to\infty} C(u_{n_k}) = \inf_{u\in U} C(u),$$

that is, u^* attains the infimum.

The case of U bounded follows similarly, since now any minimizing sequence is automatically bounded in H.

To prove uniqueness, we note that the cost C is actually *strictly convex*, that is,

$$C(\tfrac{1}{2}u_1 + \tfrac{1}{2}u_2) < \tfrac{1}{2}C(u_1) + \tfrac{1}{2}C(u_2)$$

if $u_1 \neq u_2$.

Then if u_1 and u_2 both minimize C over U,

$$C(\tfrac{1}{2}u_1 + \tfrac{1}{2}u_2) < \tfrac{1}{2}C(u_1) + \tfrac{1}{2}C(u_2) = \inf_{u\in U} C(u)$$

by strict convexity. However, since U is a convex set $\tfrac{1}{2}u_1 + \tfrac{1}{2}u_2 \in U$, a contradiction, unless $u_1 = u_2$. ■

The next theorem is the direct analog of the directional derivative test for a critical point from Section 1.

Theorem 2 An element $u^* \in U$ minimizes C over U if and only if

$$a(u^*, v) - a(u^*, u^*) \geq L(v) - L(u^*) \qquad \text{for all} \quad v \in U. \tag{2}$$

Proof Suppose u^* is a minimizing element. Then for any $v \in U$ and $0 \leq \theta \leq 1$, $(1-\theta)u^* + \theta v \in U$; hence

$$C(u^*) \leq C((1-\theta)u^* + \theta v),$$

whence

$$\frac{1}{\theta}[C(u^* + \theta(v - u^*)) - C(u^*)] \geq 0.$$

Substituting for C and using the bilinearity of a and the linearity of L gives in the limit as $\theta \to 0$

$$a(u^*, v) - a(u^*, u^*) \geq L(v) - L(u^*) \qquad \text{for all} \quad v \in U,$$

as required.

Conversely, suppose (2) holds. Since C is convex,

$$C((1-\theta)w+\theta v) \le (1-\theta)C(w)+\theta C(v)$$
$$= C(w)+\theta(C(v)-C(w));$$

hence

$$C(v)-C(w) \ge \frac{1}{\theta}[C((1-\theta)w+\theta v)-C(w)].$$

Now expanding and letting $\theta \to 0$, we find as before that

$$C(v)-C(w) \ge a(w,v)-a(w,w)-L(v)+L(w).$$

Setting $w = u^*$ gives

$$C(v)-C(u^*) \ge a(u^*,v)-a(u^*,u^*)-L(v)+L(u^*) \ge 0 \quad \text{for all} \quad v \in U,$$

by hypothesis. Consequently, $C(v) \ge C(u^*)$ for all $v \in U$; hence u^* is a minimizing element. ■

Inequalities of the type (2) in Theorem 2 are called *variational inequalities.*

A special case of this theorem is worth noticing.

Corollary 1 In the case $U = H$ (that is, there are no control constraints), (2) becomes

$$a(u^*,v) = L(v) \qquad \text{for all} \quad v \in H. \tag{2'}$$

Proof Set $v = u^* \pm \varphi$ in (2), where φ is any element in U. Then $v \in U$; hence

$$a(u^*,v)-a(u^*,u^*) \ge L(v)-L(u^*),$$
$$a(u^*,\varphi) \ge L(\varphi) \qquad \text{when} \quad v = u^*+\varphi, \tag{3}$$
$$a(u^*,-\varphi) \ge L(-\varphi) \qquad \text{when} \quad v = u^*-\varphi,$$

that is,

$$a(u^*,\varphi) \le L(\varphi),$$

and equality in (3) holds. ■

Corollary 1 is a special case of the well-known Lax–Milgram theorem.

The application of Theorems 1 and 2 to control problems of the type in Section 1 follows since

$$C(u) = \int_\Omega (y(u) - y_d)^2 \, d\mathbf{x} = \int_\Omega y(u)^2 \, d\mathbf{x} - 2\int_\Omega y(u) y_d \, d\mathbf{x} + \int_\Omega y_d^2 \, d\mathbf{x}$$
$$= (y(u), y(u)) - 2(y(u), y_d) + (y_d, y_d),$$

where $(u, v) = \int_\Omega uv \, d\mathbf{x}$ is the $L^2(\Omega)$ inner product. Since y_d is fixed, C is minimized at the same time as

$$(y(u), y(u)) - 2(y(u), y_d),$$

which is in the general form considered in Theorems 1 and 2 with

$$a(u, u) = (y(u), y(u)), \qquad u \in H = L^2(\Omega),$$
$$L(u) = (y(u), y_d).$$

By Theorem 2, the necessary condition for optimality given in Section 1 is in fact sufficient, and the existence and uniqueness of an optimal control can be derived from Theorem 1. [Note that in the case $f \neq 0$, the mapping $u \to y(u)$ is only affine, not linear; however, it is easily seen from the proofs that the theorems still hold.]

Theorems 1 and 2 and in particular Corollary 1 also have immediate application to the existence and uniqueness of weak solutions to partial differential equations.

Example 1 Let

$$U = H = H^1(\Omega) = \left\{ v : v, \frac{\partial v}{\partial x_i} \in L^2(\Omega), \qquad \text{for} \quad i = 1, 2, \ldots, n \right\}.$$

$H^1(\Omega)$ is a Hilbert space with norm

$$\|v\|_1^2 = \int_\Omega \left(\sum_{i=1}^n \left| \frac{\partial v}{\partial x_i} \right|^2 + |v|^2 \right) d\mathbf{x}.$$

Define, for $u, v \in H^1(\Omega)$,

$$a(u, v) = \int_\Omega (\nabla u \cdot \nabla v + uv) \, d\mathbf{x}$$

and

$$L(v) = \int_\Omega fv \, d\mathbf{x}$$

for some fixed element $f \in L^2(\Omega)$. It is not difficult to show that a is a symmetric, bilinear, bounded form on H, and L is continuous and linear. The coercivity of a follows directly since

$$a(u, u) = \int (|\nabla u|^2 + u^2)\, d\mathbf{x} = \|u\|_1^2.$$

Consequently, by Theorem 1 and Corollary 1, there exists a unique $u^* \in H^1(\Omega)$ satisfying

$$a(u^*, v) = L(v) \qquad \text{for all} \quad v \in H^1(\Omega),$$

or that

$$\int_\Omega (\nabla u^* \cdot \nabla v + u^* v)\, d\mathbf{x} = \int_\Omega fv\, d\mathbf{x} \quad \text{for all} \quad v \in H^1(\Omega). \tag{4}$$

Thus u^* is the unique weak solution of the elliptic equation

$$\begin{aligned} -\Delta u + u &= f \qquad \text{in} \quad \Omega, \\ \frac{\partial u}{\partial n} &= 0 \qquad \text{on} \quad \Gamma. \end{aligned} \tag{5}$$

The correspondence between solutions of (4) and (5) can be obtained by Green's identity. Namely, if u satisfies (5), then, for any $v \in H^1(\Omega)$,

$$\begin{aligned} \int_\Omega fv\, d\mathbf{x} &= -\int_\Omega v(\Delta u - u)\, d\mathbf{x} = -\int_\Gamma v \frac{\partial u}{\partial n} + \int_\Omega \nabla u \cdot \nabla v\, d\mathbf{x} + \int_\Omega uv\, d\mathbf{x} \\ &= \int_\Omega (\nabla u \cdot \nabla v + uv)\, d\mathbf{x}. \end{aligned}$$

Consequently, classical solutions of (5) obey (4). However, a priori solutions of (4) need not even have two derivatives; hence the term weak solution.

Example 2 If we instead choose $U = H_0^1(\Omega)$ [the subspace of elements in $H^1(\Omega)$ vanishing on Γ], then choosing a and L as before, we find a unique $u^* \in H_0^1(\Omega)$ satisfying

$$\int_\Omega (\nabla u^* \cdot \nabla v + u^* v)\, d\mathbf{x} = \int_\Omega fv\, d\mathbf{x} \qquad \text{for all} \quad v \in H_0^1(\Omega).$$

Another application of Green's identity shows that u^* is the weak solution of the Dirichlet problem

$$\begin{aligned} -\Delta u + u &= f \qquad \text{in} \quad \Omega, \\ u &= 0 \qquad \text{on} \quad \Gamma. \end{aligned}$$

3. BOUNDARY CONTROL AND APPROXIMATE CONTROLLABILITY OF ELLIPTIC SYSTEMS

Boundary Control

We reconsider the control problem in Section 1 except now the control will be exercised through the boundary. Namely, suppose for a given control $u \in L^2(\Gamma)$ that $y(u)$ is the solution of the Dirichlet problem

$$\begin{aligned} \Delta y(u) &= 0 \quad \text{in} \quad \Omega, \\ y(u) &= u \quad \text{on} \quad \Gamma, \end{aligned} \tag{1}$$

Again a desired state $y_{\mathrm{d}} \in L^2(\Omega)$ is given, and we wish to minimize

$$C(u) = \int_\Omega (y(u) - y_{\mathrm{d}})^2 \, d\mathbf{x},$$

over the set of admissible controls U, a bounded, closed, convex subset of $L^2(\Gamma)$. The necessary and sufficient condition for optimality becomes, as before,

$$\int_\Omega (y(u^*) - y_{\mathrm{d}})(y(v) - y(u^*)) \, d\mathbf{x} \geq 0 \qquad \text{for all} \quad v \in U. \tag{2}$$

If we introduce the adjoint system

$$\begin{aligned} \Delta p(u) &= y(u) - y_{\mathrm{d}} \quad \text{in} \quad \Omega, \\ p(u) &= 0 \quad \text{on} \quad \Gamma. \end{aligned} \tag{3}$$

Then for $u \in L^2(\Gamma)$, we have, by Green's identity,

$$\begin{aligned} \int_\Omega \Delta p(u^*) y(v - u^*) \, d\mathbf{x} = &-\int_\Gamma \frac{\partial p(u^*)}{\partial n} y(v - u^*) + \int_\Gamma p(u^*) \frac{\partial y(v - u^*)}{\partial n} \\ &+ \int_\Omega p(u^*) \, \Delta y(v - u^*) \, d\mathbf{x}. \end{aligned} \tag{4}$$

Using the state and adjoint equations, (4) becomes

$$\int_\Omega \Delta p(u^*) y(v - u^*) \, d\mathbf{x} = -\int_\Gamma \frac{\partial p(u^*)}{\partial n} (v - u^*), \tag{5}$$

and hence the necessary and sufficient condition for optimality is now

$$\int_\Gamma \frac{\partial p(u^*)}{\partial n} (v - u^*) \leq 0 \qquad \text{for all} \quad v \in U. \tag{6}$$

Again, in the case $U = \{v: |v(x)| \leq 1, x \in \Gamma\}$, (6) can be shown to be equivalent to the local condition

$$\frac{\partial p(u^*)}{\partial n}(\mathbf{x})v(\mathbf{x}) \leq \frac{\partial p(u^*)}{\partial n}(\mathbf{x})u^*(\mathbf{x}) \qquad \text{a.e.} \quad \mathbf{x} \in \Gamma,$$

that is,

$$u^*(\mathbf{x}) = \operatorname{sgn}\left\{\frac{\partial p(u^*)}{\partial n}(\mathbf{x})\right\}, \qquad \text{a.e.} \quad \mathbf{x} \in \Gamma. \tag{7}$$

Remark In the case of Neumann boundary control,

$$\begin{aligned} \Delta y(u) &= f \qquad \text{in} \quad \Omega, \\ \frac{\partial y(u)}{\partial n} &= u \qquad \text{on} \quad \Gamma, \end{aligned}$$

the optimality condition becomes

$$\int_\Gamma p(u^*)(v - u^*) \geq 0 \qquad \text{for all} \quad v \in U, \tag{8}$$

with the adjoint system

$$\begin{aligned} \Delta p(u) &= y(u) - y_{\mathrm{d}} \qquad \text{in} \quad \Omega, \\ \frac{\partial p(u)}{\partial n} &= 0 \qquad \text{on} \quad \Gamma. \end{aligned}$$

The analog of (7) is now

$$u^* = -\operatorname{sgn}\{p(u^*)\}.$$

Approximate Controllability

In Section 1 we supposed a state $y_{\mathrm{d}} \in L^2(\Omega)$ was given, and we tried to minimize the cost

$$C(u) = \int_\Omega (y(u) - y_{\mathrm{d}})^2 \, d\mathbf{x}$$

subject to

$$\begin{aligned} \Delta y(u) &= u \qquad \text{in} \quad \Omega, \\ y(u) &= 0 \qquad \text{on} \quad \Gamma. \end{aligned} \tag{9}$$

A natural question to ask is, if the controls are allowed to vary over $L^2(\Omega)$, can any desired state y_d be approximated as closely as we like by some $y(u)$, for $u \in L^2(\Omega)$? In other words, is the set of states $\{y(u): u \in L^2(\Omega)\}$ dense in $L^2(\Omega)$. Of course this is a controllability problem. Now instead of trying to reach a desired state exactly [in $L^2(\Omega)$ this is clearly ruled out by the regularity of $y(u)$], we ask if a desired state can be approximated as closely as we like by controls in $L^2(\Omega)$. This concept is called *approximate controllability*.

It is a well-known consequence of the Hahn–Banach theorem that the set $\{y(u) : u \in L^2(\Omega)\}$ is dense in $L^2(\Omega)$, and hence (9) is approximately controllable there if and only if

$$\eta \in L^2(\Omega) \qquad \text{and} \qquad \int_\Omega \eta y(u)\, d\mathbf{x} = 0 \qquad \text{for all} \quad u \in L^2(\Omega), \tag{10}$$

implies that $\eta = 0$. If we define the adjoint equation as before,

$$\begin{aligned} \Delta p &= \eta \quad \text{in} \quad \Omega, \\ p &= 0 \quad \text{on} \quad \Gamma, \end{aligned}$$

then

$$\int_\Omega \eta y(u)\, d\mathbf{x} = \int_\Omega \Delta p y(u)\, d\mathbf{x} = \int_\Omega pu\, d\mathbf{x}.$$

Hence (10) implies that

$$\int_\Omega pu\, d\mathbf{x} = 0 \qquad \text{for all} \quad u \in L^2(\Omega),$$

and so

$$p \equiv 0,$$

that is,

$$\Delta p = \eta \equiv 0,$$

and the system (9) is approximately controllable.

In the case of boundary control, however, the system is not approximately controllable in $L^2(\Omega)$. For now suppose

$$\int_\Omega \eta(\mathbf{x}) y(u)(\mathbf{x})\, d\mathbf{x} = 0 \qquad \text{for} \quad \eta \in L^2(\Omega),$$

where

$$\Delta y(u) = 0 \quad \text{in} \quad \Omega,$$
$$y(u) = u \quad \text{on} \quad \Gamma.$$

If we define the adjoint system by

$$\Delta p = \eta \quad \text{in} \quad \Omega,$$
$$p = 0 \quad \text{on} \quad \Gamma,$$

then

$$\begin{aligned} 0 &= \int_\Omega \eta y(u)\, d\mathbf{x} = \int_\Omega \Delta p y(u)\, d\mathbf{x} \\ &= -\int_\Gamma \frac{\partial p}{\partial n} y(u) + \int_\Gamma p \frac{\partial y(u)}{\partial n} + \int_\Omega p \Delta y(u)\, d\mathbf{x} \\ &= -\int_\Gamma \frac{\partial p}{\partial n} u. \end{aligned}$$

Consequently, $\int_\Gamma (\partial p/\partial n) u = 0$ for all $u \in L^2(\Gamma)$; hence $\partial p/\partial n = 0$ on Γ. Thus the adjoint system becomes

$$\begin{aligned} \Delta p &= \eta \quad \text{in} \quad \Omega, \\ p &= 0 \quad \text{on} \quad \Gamma, \\ \frac{\partial p}{\partial n} &= 0 \quad \text{on} \quad \Gamma. \end{aligned} \tag{11}$$

However, it is now possible to find a nonzero $\eta \in L^2(\Omega)$ satisfying (11). For instance, if $n = 1$, we can choose p as shown in Fig. 2 and pick $\eta = p''$. For such an η, we have

$$\int_\Omega \eta y(u) = 0 \qquad \text{for all} \quad u \in L^2(\Gamma),$$

and so $\{y(u) : u \in L^2(\Gamma)\}$ is not dense in $L^2(\Omega)$ and the system is not approximately controllable.

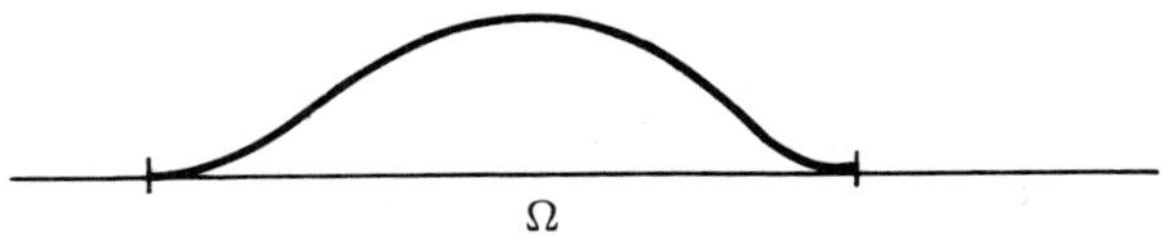

Fig. 2

4. THE CONTROL OF SYSTEMS GOVERNED BY PARABOLIC EQUATIONS

In this section we consider certain control systems governed by parabolic partial differential equations that typically arise in heat conduction or diffusion problems [1]. For example, if the temperature in a homogeneous body Ω (Fig. 3) is controlled by a variable heat source $u(\mathbf{x}, t)$ for $\mathbf{x} \in \Omega$, $t > 0$; if we denote the temperature corresponding to u by $y(\mathbf{x}, t : u)$ or just $y(t : u)$ or $y(u)$; and if the boundary of Ω is kept at a constant, say zero, temperature, then y satisfies

$$\begin{aligned} \frac{\partial y(u)}{\partial t} &= c\,\Delta y(u) + f + u && \text{in} \quad \Omega, \\ y(\mathbf{x}, 0 : u) &= y_0(\mathbf{x}), && \mathbf{x} \in \Omega, \\ y(u) &= 0 && \text{on} \quad \Gamma. \end{aligned} \tag{1}$$

The initial temperature y_0 and fixed source term f will be assumed to be elements of $L^2(\Omega)$. For simplicity, we take $c = 1$. [The case of insulated boundary $\partial y(u)/\partial n = 0$ can be handled analogously.] If the controls u are constrained to belong to a given set of admissible controls $U \subset L^2(\Omega)$, and a desired temperature distribution $y_\text{d} \in L^2(\Omega)$ is given at some final, fixed time T, then one control problem of interest is to choose an admissible control u that brings $y(u)$ as close to y_d as possible in the given time T. That is,

$$\min_{u \in U} C(u) = \int_\Omega (y(T : u) - y_\text{d})^2 \, d\mathbf{x},$$

where we again use the mean-square-error criterion. Since the cost C is essentially the same type as before, our previous theory applies. In particular, if U is a bounded, closed, convex set in $L^2(\Omega)$, then an optimal control u^* exists. The necessary and sufficient condition for

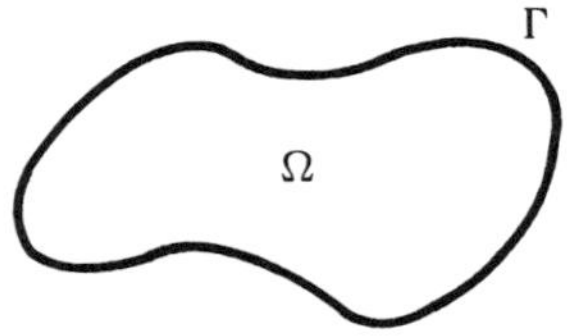

Fig. 3

optimality is

$$\int_\Omega (y(T:u^*) - y_d)(y(T:v) - y(T:u^*))\,d\mathbf{x} \geq 0 \qquad \text{for all} \quad v \in U. \qquad (2)$$

Introducing the adjoint system,

$$\begin{aligned} \frac{\partial p(u)}{\partial t} + \Delta p(u) &= 0 && \text{in} \quad \Omega, \\ p(T:u) &= y(T:u) - y_d && \text{in} \quad \Omega, \\ p(u) &= 0 && \text{on} \quad \partial\Omega, \end{aligned}$$

we have that

$$\begin{aligned} 0 &= \int_0^T \int_\Omega \left\{ \frac{\partial p(u^*)}{\partial t} + \Delta p(u^*) \right\} (y(t:v) - y(t:u^*))\,d\mathbf{x}\,dt \\ &= \int_\Omega \int_0^T \frac{\partial p(^*)}{\partial t} (y(t:v) - y(t:u^*))\,d\mathbf{x}\,dt \\ &\quad + \int_0^T \int_\Omega \Delta p(u^*)(y(t:v) - y(t:u^*))\,d\mathbf{x}\,dt. \end{aligned}$$

On integrating by parts,

$$\begin{aligned} 0 &= \int_\Omega \left\{ p(t:u^*)(y(t:v) - y(t:u^*))\Big|_0^T \right. \\ &\quad \left. - \int_0^T p(u^*)\left(\frac{\partial y(v)}{\partial t} - \frac{\partial y(u^*)}{\partial t}\right) dt \right\} d\mathbf{x} \\ &\quad + \int_0^T \int_\Omega p(u^*)(\Delta y(v) - \Delta y(u^*))\,d\mathbf{x}\,dt \\ &= \int_\Omega p(T:u^*)(y(T:v) - y(T:u^*))\,d\mathbf{x} \\ &\quad - \int_0^T \int_\Omega p(u^*)\left(\frac{\partial y(v)}{\partial t} - \Delta y(v) - \frac{\partial y(u^*)}{\partial t} + \Delta y(u^*)\right) d\mathbf{x}\,dt \\ &= \int_0^T p(T:u^*)(y(T:v) - y(T:u^*))\,d\mathbf{x} - \int_0^T \int_\Omega p(u^*)(v - u^*)\,d\mathbf{x}\,dt. \end{aligned}$$

So the necessary condition (2) becomes

$$\int_0^T \int_\Omega p(u^*)(v - u^*)\,d\mathbf{x}\,dt \geq 0 \qquad \text{for all} \quad v \in U. \qquad (3)$$

If $U = \{u : |u| \leq 1\}$, then (3) becomes

$$u^*(\mathbf{x}, t) = -\operatorname{sgn}\{p(\mathbf{x}, t : u^*)\},$$

and in this case, the optimal control u^* is completely determined by the necessary condition.

Proposition 1 [5] If $y(u^*) \neq y_d$, then the optimal control u^* is bang-bang, $|u^*| \equiv 1$, and u^* is unique.

Proof As usual, the only way u^* cannot be bang-bang and unique is if $p(\mathbf{x}, t : u^*) = 0$ for $(\mathbf{x}, t) \in E \subset \Omega \times (0, T)$, E nonnull. By taking E smaller, if necessary, we can suppose that $E \subset \Omega \times (0, t_1)$ for any $t_1 < T$ that is sufficiently close to T. Since p satisfies the heat equation with homogeneous boundary conditions, it must be analytic in $\Omega \times (0, t_1)$ [2], and hence identically zero there. Now letting $t_1 \uparrow T$ and using the continuity of $p(u^*)$ in time gives

$$p(T: u^*) = 0 \qquad \text{a.e. in} \quad \Omega,$$

and hence

$$y(u^*) = y_d,$$

a contradiction. ■

Boundary Control of the Heat Equation

We now consider the problem of controlling the temperature in the region Ω when the body is heated by convection on its boundary. For this problem, Newton's law of heating (cooling) is a good approximation, and we can write the state equations

$$\begin{aligned} y_t &= \Delta y && \text{in} \quad \Omega \\ y(\mathbf{x}, 0) &= y_0(\mathbf{x}), && \mathbf{x} \in \Omega \\ \frac{\partial y}{\partial n}(\mathbf{x}, t) &= b(u(\mathbf{x}, t) - y(\mathbf{x}, t)), && \mathbf{x} \in \Gamma, \quad t > 0, \end{aligned} \tag{4}$$

where u is the (boundary) control function and b a positive constant. In this case the set of admissible controls U will be taken as a closed, convex, bounded subset of $L^2(\Gamma \times (0, T))$. In the same manner as before, we define the adjoint system

$$\begin{aligned} \frac{\partial p(u)}{\partial t} &= \Delta p(u) && \text{in} \quad \Omega, \\ p(T: u) &= y(T: u) - y_d && \text{in} \quad \Omega, \\ \frac{\partial p(u)}{\partial n} + bp(u) &= 0 && \text{on} \quad \Gamma, \end{aligned} \tag{5}$$

and the necessary and sufficient condition for optimality becomes

$$\int_0^T \int_\Gamma p(t : u^*)(v - u^*) \geq 0 \qquad \text{for all} \quad v \in U.$$

Now in the case $U = \{u \in L^2(\Gamma \times (0, T)) : |u| \leq 1\}$, we have

$$u^*(\mathbf{x}, t) = \operatorname{sgn}(p(\mathbf{x}, t : u^*))$$

and the following proposition.

Proposition 2 (Knowles [3]; Lions [5]) If $y(u^*) \neq y_d$ and Γ is analytic, then u^* is bang-bang and unique.

Proof (Sketch) Under the above assumptions, $p(u^*)$ is analytic in $\bar{\Omega} \times (0, t_1)$, $t_1 < T$[2]. Consequently, if $p(u^*)$ vanishes on a nonnull set in $\Gamma \times (0, t_1)$, it must be identically zero on $\Gamma \times (0, t_1)$, and hence so must $\partial p(u^*)/\partial n$ by (5). Since $p(u^*)$ is analytic in $\Omega \times (0, t_1)$ and since $p(u^*)$ and $\partial p(u^*)/\partial n$ vanish on $\Gamma \times (0, t_1)$, uniqueness requires that $p(u^*) \equiv 0$ in $\bar{\Omega} \times (0, t_1)$. As before, by continuity, this implies $y_d = y(T: u^*)$, a contradiction. ∎

Corollary 1 In the case in which $n = 1$ and Ω is interval, the optimal control $u^* = u^*(t)$ can, under the assumptions of Theorem 2, have at most a countable number of switches, and if so, the switching times must accumulate at T.

Proof $u^*(t) = \operatorname{sgn}\{p(t : u^*)\}$, and by analyticity, p can have only a finite number of zeros in any interval $(0, t_1)$, $t_1 < T$. ∎

5. TIME OPTIMAL CONTROL

Instead of taking the terminal time T fixed, we now consider certain extensions of the time optimal control problem of Chapter I to systems governed by parabolic partial differential equations. As we noted in our earlier remarks on approximate controllability (see also Section 6), the *exact hitting* of the desired final state y_d may not be possible in *any finite time*, although we may be able to approach y_d as closely as we like in finite time. Consequently, a more practical formulation of the time optimal control problem for partial differential equations, is to suppose a "terminal error" ε is preassigned, and to try to hit the *target set*

$$W = B(y_d, \varepsilon) = \{y : \|y - y_d\| \leq \varepsilon\} \tag{1}$$

in minimum time, using admissible controls. Since we have been assuming $y_d \in L^2(\Omega)$, we shall take the norm in (1) to be the $L^2(\Omega)$ norm. [Other choices of the norm in (1) are clearly possible and may, in fact, be more interesting in applications.]

If $y(t:u)$ is the solution of the state equations (4.1) at time t [or alternatively (4.4)] corresponding to a control $u \in U$, a closed, convex, bounded subset of $L^2(\Omega)$ [respectively $L^2(\Gamma)$], then we seek a control $u^* \in U$ such that $y(t^*:u^*) \in W$ and t^* is minimal. Given that *some* admissible control reaches W in finite time, it can be shown by basically the same geometric methods as used in Chapter I that the minimum time t^* exists [3, 5].

A necessary condition characterizing t^* and u^*, the optimal control, can be obtained from the previous problem. Namely, in the control problem with cost

$$C(u) = \|y(t^*:u) - y_d\|, \tag{2}$$

u^* must also be optimal, with minimal cost ε; that is,

$$C(u^*) = \min_{u \in U} C(u) = \varepsilon.$$

For if $\bar{u} \in U$ achieves a smaller cost in (2),

$$C(\bar{u}) < \varepsilon,$$

then

$$\|y(t^*:\bar{u}) - y_d\| < \varepsilon.$$

However, the mapping $t \mapsto y(t:\bar{u})$, $[0, T] \to L^2(\Omega)$, is continuous, so we must be able to find a smaller time $\bar{t} < t^*$ with

$$\|y(\bar{t}:\bar{u}) - y_d\| \leq \varepsilon,$$

which contradicts the minimality of t^*. Conversely, it can be similarly argued that if u_1 minimizes the cost (2), then u_1 must also be a time optimal control.

By virtue of this equivalence, the optimality conditions can be deduced from Section 4. For the time optimal control problem with state equations (4.1) and corresponding adjoint system, the optimality conditions may be stated as

$$\int_0^{t^*} \int_\Omega p(u^*)(v - u^*)\, d\mathbf{x}\, dt \geq 0 \qquad \text{for all} \quad v \in U. \tag{3}$$

For the boundary control problem (4.4), with adjoint equations (4.5), the optimality conditions now are

$$\int_0^{t^*} \int_\Gamma p(u^*)(v - u^*) \geq 0 \qquad \text{for all} \quad v \in U. \tag{4}$$

The analogous local necessary conditions also hold, as in fact do the conclusions of Proposition 4.2 and Corollary 4.1.

Proposition 1 (Knowles [3]) If the set of admissible controls is $U = \{u \in L^2(\Omega \times [0, T]) : |u(\mathbf{x})| \le 1\}$ for the problem of Eqs. (4.1) or $U = \{u \in L^2(\Gamma \times [0, T]) : |u(\mathbf{x})| \le 1\}$ for the problem of Eqs. (4.4) and Γ is analytic, then time-optimal controls for both problems are unique, bang-bang ($|u(\mathbf{x})| \equiv 1$), and uniquely determined by (3) [respectively (4)].

6. APPROXIMATE CONTROLLABILITY FOR PARABOLIC PROBLEMS

In this section we give a partial answer to the controllability questions raised in Section 5. Namely, if $y(t : u)$ denotes the solution of (4.4) for $u \in L^2(\Gamma \times [0, T])$, then we shall show that $\{y(T: u) : u \in L^2(\Gamma \times [0, T])\}$ is dense in $L^2(\Omega)$, *for any* $T > 0$, and so the boundary control problem (4.4) is approximately controllable. Again, we cannot possibly expect exact controllability in $L^2(\Omega)$ (or $H^p(\Omega)$ for that matter), because of the smoothing properties of parabolic differential equations. By the same methods one can show the distributed control problem of Eqs. (4.1) is also approximately controllable in $L^2(\Omega)$ in any time $T > 0$.

If $\{y(T: u) : u \in L^2(\Gamma \times [0, T])\}$ were not dense in $L^2(\Omega)$, then, by the Hahn–Banach theorem, we could find a nonzero $\eta \in L^2(\Omega)$ such that

$$\int_\Omega \eta y(T : u)\, d\mathbf{x} = 0 \qquad \text{for all} \quad u \in L^2(\Gamma \times [0, T]). \tag{1}$$

Introducing the adjoint system

$$\begin{aligned} \frac{\partial p}{\partial t} + \Delta p &= 0 && \text{in} \quad \Omega \times (0, T), \\ \frac{\partial p}{\partial n} + bp &= 0 && \text{on} \quad \Gamma \times (0, T), \\ p(\mathbf{x}, T) &= \eta(\mathbf{x}), && \mathbf{x} \in \Omega, \end{aligned}$$

(1) then implies that

$$\int_\Omega p(T) y(T : u)\, d\mathbf{x} = 0 \qquad \text{for all} \quad u \in L^2(\Gamma \times [0, T]), \tag{2}$$

and on integration by parts, (2) leads to

$$\int_0^T \int_\Gamma pu = 0 \qquad \text{for all} \quad u \in L^2(\Gamma \times [0, T]).$$

Thus

$$p = 0 \qquad \text{on} \quad \Gamma \times (0, T),$$

and consequently

$$\frac{\partial p}{\partial n} = 0 \qquad \text{on} \quad \Gamma \times (0, T)$$

[by (4.4)]. Uniqueness then requires that $p = 0$ in $\Omega \times (0, T)$. Again taking $t \uparrow T$, and using the continuity of $t \to p(t)$, we have that $p(T) = \eta = 0$, a contradiction.

We note in passing that the similarity between the above proof and that of Propositions 4.1 and 4.2 is not coincidental. In fact, in the propositions we showed that the system had an analog of the "normality property" of Chapter I, namely. optimal controls were bang-bang, unique, and uniquely determined by the necessary condition. As in the case of ordinary differential equations, it can be shown that normality implies approximate controllability, and, in fact, is in some cases equivalent to it [4].

An analog of the distributed control problem (4.1) is to consider the action of m independently acting "point-source" controllers. In one dimension, we could consider controls acting on Ω as shown in Fig. 4. The state equation would then become

$$\begin{aligned} \frac{\partial y}{\partial t} &= \Delta y + \sum_{i=1}^{m} g_i(x) u_i(t), \qquad && x \in \Omega, \quad t > 0, \\ y(x, t) &= 0, && x \in \Gamma, \quad t > 0, \\ y(x, 0) &= 0, && x \in \Omega, \end{aligned} \tag{3}$$

with $g_i = \delta_{x_i}$, the Dirac δ function centered at x_i, $i = 1, 2, \ldots, m$. Alternatively, with $m = 1$, $g_1(x) \equiv 1$, (3) would model the problem of controlling the temperature on the rod Ω by the ambient temperature $u_1(t)$.

If λ_j and φ_j, j $= 1, 2, \ldots$, are the eigenvalues and eigenfunctions of the system $y'' = \lambda y$ in Ω with homogeneous Dirichlet boundary condi-

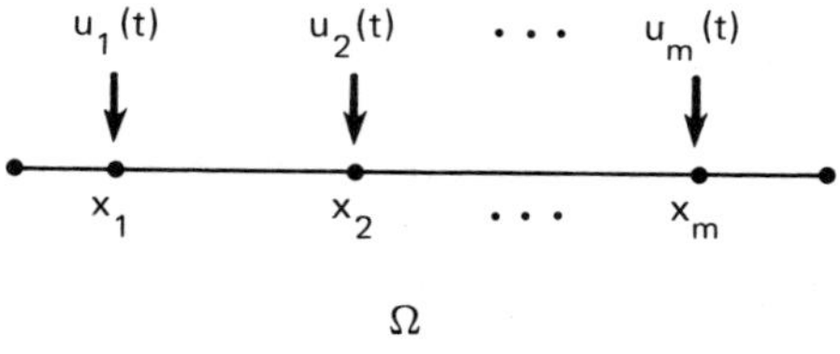

tions, then for $g_i \in L^2(\Omega)$, the solution of (3) can be written

$$y(x,t:u) = \sum_{j=1}^{\infty}\left(\sum_{i=1}^{m}(g_i,\varphi_j)\int_0^t \exp[-\lambda_j(t-\tau)]u_i(\tau)\,d\tau\right)\varphi_j(x), \qquad (4)$$

where $(g_i, \varphi_j) = \int_\Omega g_i(x)\varphi_j(x)\,dx$ are the Fourier coefficients of the $\{g_i\}$ with respect to the $\{\varphi_j\}$. The solution (4) is also valid for $g_i = \delta_{x_i}$ if we interpret $(g_i, \varphi_j) = \varphi_j(x_i)$.

As before, (3) is approximately controllable in time T if and only if

$$\int_\Omega \eta y(T:u)\,d\mathbf{x} = 0 \qquad \text{for all} \quad u = (u_i), \quad u_i \in L^2(0,T), \quad i = 1,2,\ldots,m \qquad (5)$$

implies that $\eta = 0$ ($\eta \in L^2(\Omega)$). If $m = 1$, (5) is equivalent to

$$\sum_{j=1}^{\infty}(g,\varphi_j)\left(\int_0^T \exp[-\lambda_j(t-\tau)]u(\tau)\,d\tau\right)(\eta,\varphi_j) = 0 \quad \text{for all} \quad u \in L^2(0,T),$$

and hence

$$\sum_{j=1}^{\infty}(g,\varphi_j)(\eta,\varphi_j)\exp(-\lambda_j\tau) \equiv 0 \qquad \text{for} \quad 0 < \tau \le T. \qquad (6)$$

However, the functions $\{\exp(-\lambda_j\tau)\}$ are linearly independent, and so (6) must imply that

$$(g,\varphi_j)(\eta,\varphi_j) = 0 \qquad \text{for all} \quad j = 1,2,\ldots. \qquad (7)$$

The necessary and sufficient condition for approximate controllability can now be read from (7). If $(g,\varphi_{j_0}) = 0$ for some j_0, then we can certainly find a nonzero η satisfying (7); for instance, take $\eta = \varphi_j$ for any $j \ne j_0$.

Conversely, if $(g,\varphi_j) \ne 0$ for all $j = 1,2,\ldots$, then from (7) $(\eta,\varphi_j) = 0$ for all j, and hence $\eta = 0$, since the $\{\varphi_j\}$ span $L^2(\Omega)$. *Consequently, with* $m = 1$, (3) *is approximately controllable if and only if* $(g,\varphi_j) \ne 0$ *for every* $j = 1,2,\ldots$.

In the general case, with m controls one can similarly show (5) is equivalent to

$$(g_i,\varphi_j)(\eta,\varphi_j) = 0 \qquad \text{for all} \quad i = 1,2,\ldots,m, \quad j = 1,2,\ldots,m.$$

Consequently, (4.4) *is approximately controllable in any time* $T > 0$ *if and only if*

$$\text{for every} \quad j = 1,2,\ldots, \qquad (g_i,\varphi_j) \ne 0 \quad \text{for some} \quad i = 1,2,\ldots,m. \qquad (8)$$

Note that in the case $g_i = \delta_{x_i}$, (8) becomes

$$\text{for every} \quad j = 1,2,\ldots, \qquad \varphi_j(x_i) \ne 0 \quad \text{for some} \quad i = 1,2,\ldots,m,$$

or equivalently, for every j, at least one x_i is not a nodal point of φ_j, $i = 1, 2, \ldots, m$.

Finally, in the case in which the state equations (3) are over a region $\Omega \subset R^N$, if λ_j are the distinct eigenvalues of

$$\begin{aligned} \Delta y &= \lambda y \qquad \text{in} \quad \Omega, \\ y &= 0 \qquad \text{on} \quad \Gamma, \end{aligned}$$

and $\{\varphi_{k,j}\}_{k=1}^{m_j}$ are corresponding eigenfunctions (m_j is the multiplicity of λ_j), then the solution of (3) becomes

$$y(x, t : u) = \sum_{j=1}^{\infty} \sum_{k=1}^{m_j} \sum_{i=1}^{m} \left(\int_0^t \exp[-\lambda_j(t - \tau)] u_i(\tau)\, d\tau \right) (g_i, \varphi_{kj}) \varphi_{kj}(x)$$

and the same argument now shows (3) *is approximately controllable in any time T if and only if*

$$\operatorname{rank} \begin{bmatrix} (g_1, \varphi_{1j}) & (g_2, \varphi_{1j}) & \cdots & (g_m, \varphi_{1j}) \\ (g_1, \varphi_{2j}) & (g_2, \varphi_{2j}) & \cdots & (g_m, \varphi_{2j}) \\ \vdots & \vdots & & \vdots \\ (g_1, \varphi_{m_j,j}) & (g_2, \varphi_{m_j,j}) & \cdots & (g_m, \varphi_{m_j,j}) \end{bmatrix} = m_j$$

for every $j = 1, 2, \ldots$. In particular, one needs *at least as many controls as the largest multiplicity of the eigenvalues*. Consequently, any system (3) with infinite multiplicity of its eigenvalues cannot be approximately controllable in $L^2(\Omega)$ with "point controls." For further discussion of this subject, we refer the reader to Triggiani [6].

References

[1] A. Butkovskiy, "Distributed Control Systems." Elsevier, Amsterdam, 1969.

[2] A. Friedman, "Partial Differential Equations." Holt, New York, 1969.

[3] G. Knowles, Time optimal control in infinite dimensional spaces. *SIAM J. Control Optim.* **14**, 919–933 (1976).

[4] G. Knowles, Some problems in the control of distributed systems, and their numerical solution, *SIAM J. Control Optim.* **17**, 5–22 (1979).

[5] J. Lions, "Optimal Control of Systems Governed by Partial Differential Equations." Springer-Verlag, Berlin and New York, 1971.

[6] R. Triggiani, Extensions of rank conditions for controllability and observability to Banach spaces and unbounded operators, *SIAM J. Control Optim.* **14**, 313–338 (1976).

Appendix I — Geometry of R^n

Suppose that C is a set contained in R^n, and that for $\mathbf{x} \in C$, $\|\mathbf{x}\| = (x_1^2 + x_2^2 + \cdots + x_n^2)^{1/2}$.

C is called *bounded* if there exists a number $\alpha < \infty$ such that $\|\mathbf{x}\| \leq \alpha$ for all $\mathbf{x} \in C$.

C is called *closed* if it contains its boundary; i.e., if $\mathbf{x}_n \in C$ and $\|\mathbf{x}_n - \mathbf{x}\| \to 0$, then we must have $\mathbf{x} \in C$.

C is *convex* if it contains every line segment whose end points lie in C; i.e., if $\mathbf{x}, \mathbf{y} \in C$, then $\frac{1}{2}\mathbf{x} + \frac{1}{2}\mathbf{y} \in C$. (See Fig. 1.)

An element $\mathbf{x} \in C$ is called an *extreme point* of C if it does not lie on any (nontrivial) line segment contained in C (alternatively, if it cannot be written as a proper linear combination of elements of C). For example, the vertices of a square are extreme points (see Fig. 2).

Given a vector $\boldsymbol{\eta} \in R^n$, the *hyperplane* through a point $\mathbf{x}_1 \in R^n$ with normal $\boldsymbol{\eta}$ is $H = \{\mathbf{x} \in R^n : \boldsymbol{\eta}^{\mathrm{T}}\mathbf{x} = \boldsymbol{\eta}^{\mathrm{T}}\mathbf{x}_1\}$ (see Fig. 3). A point $\mathbf{x}_1 \in C$ is *supported* (Fig. 4) by this hyperplane H if C lies on one side of H, i.e., $\boldsymbol{\eta}^{\mathrm{T}}\mathbf{x}_1 \geq \boldsymbol{\eta}^{\mathrm{T}}\mathbf{x}$ for all $\mathbf{x} \in C$. The point $\mathbf{x}_1$ is called an *exposed point* of C if $\mathbf{x}_1$ is supported by a hyperplane H (with normal $\boldsymbol{\eta}$, say) and if $\mathbf{x}_1$ is the only point of C supported by H. In other words $\boldsymbol{\eta}^{\mathrm{T}}\mathbf{x}_1 > \boldsymbol{\eta}^{\mathrm{T}}\mathbf{x}$ for all $\mathbf{x} \in C$, and $\mathbf{x} \neq \mathbf{x}_1$. In Fig. 5, $\mathbf{x}_1$ is an exposed point. In Fig. 6, $\mathbf{x}_1$ is a support point but not exposed point, and in Fig. 7, $\mathbf{x}_1$ is an extreme point and a support point, but not exposed point.

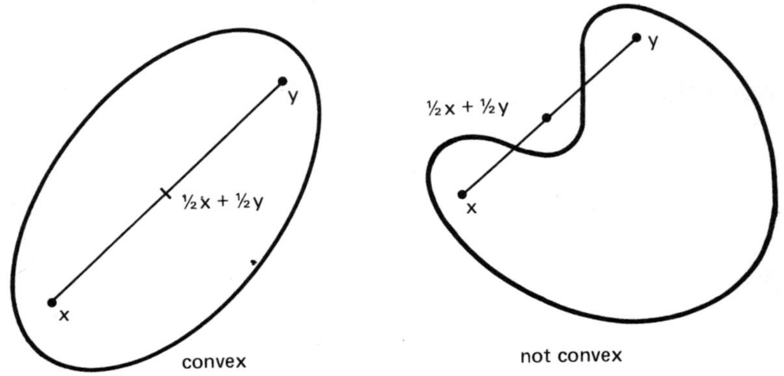

Fig. 1

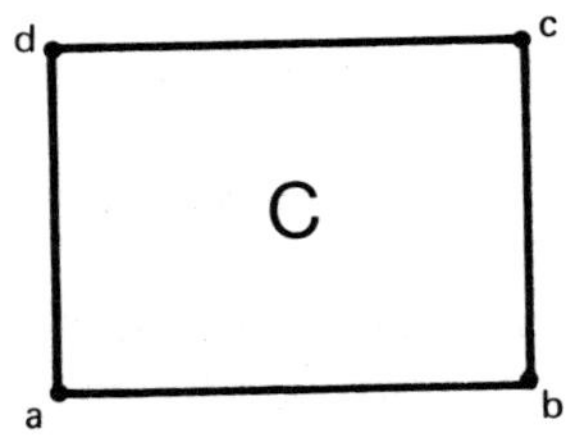

Fig. 2

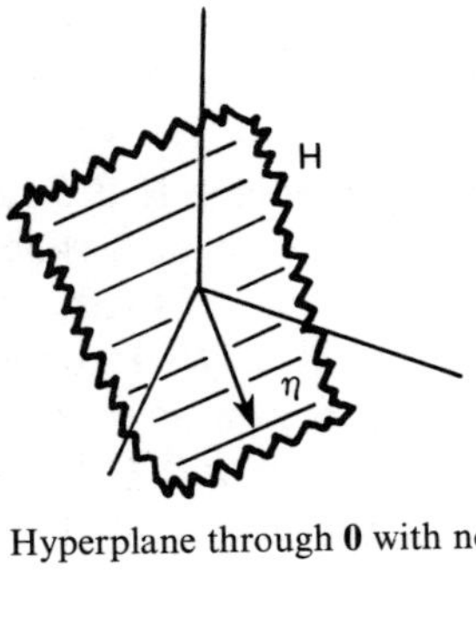

Fig. 3. Hyperplane through **0** with normal $\boldsymbol{\eta}$.

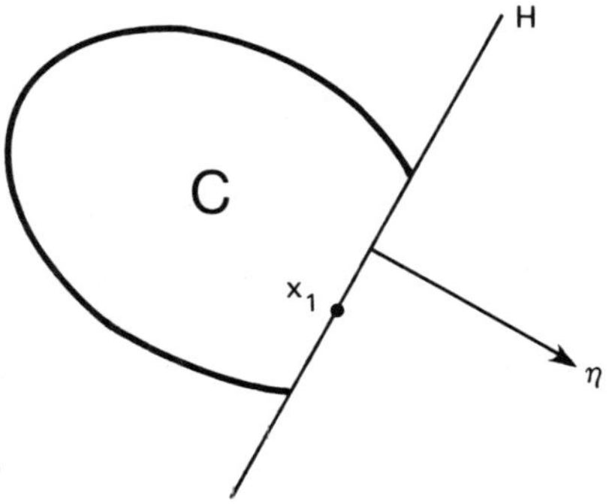

Fig. 4. $\mathbf{x}_1$ is a *support point* of C.

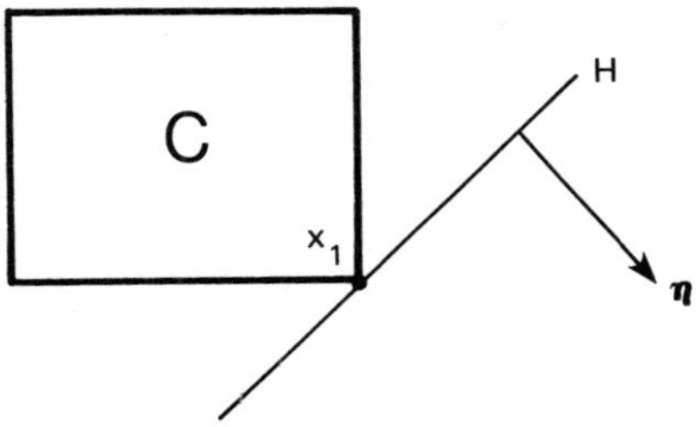

Fig. 5

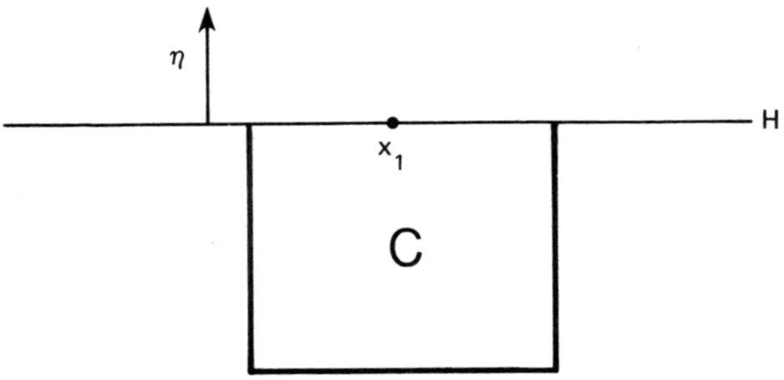

Fig. 6

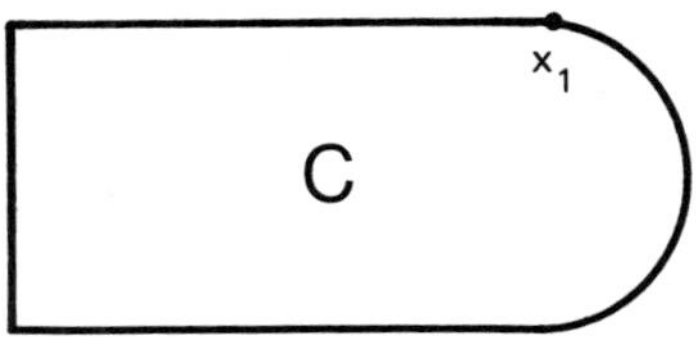

Fig. 7

Theorem 1 If C is a closed, bounded convex set, and if $\mathbf{x}_1$ belongs to the boundary of C, then we can find a hyperplane H (with normal $\boldsymbol{\eta} \neq 0$) that supports C at $\mathbf{x}_1$.

A set C is called *strictly convex* if every boundary point of C is an exposed point (e.g., circles, ellipsoids). A rectangle is not strictly convex.

In general every exposed point is a support point (trivial), and every exposed point is an extreme point (why?), but not conversely.

Appendix II — Existence of Time Optimal Controls and the Bang-Bang Principle

1. LINEAR CONTROL SYSTEMS

We shall now suppose that we are given a control system of the form

$$\dot{\mathbf{x}}(t) = A(t)\mathbf{x}(t) + \mathbf{b}(t)u(t), \qquad \mathbf{x}(0) = \mathbf{x}_0, \tag{1}$$

where $\mathbf{x}(t) \in R^n$, $u(t) \in R$, and A and $\mathbf{b}$ are $n \times n$ and $n \times 1$ matrices of functions that are integrable on any finite interval. The controls u will be restricted to the set

$$U = \{u : u \text{ is Lebesgue measurable and } |u(\tau)| \le 1 \text{ a.e.}\}, \tag{2}$$

where U is commonly called the set of *admissible controls*; note that $U \subset L^\infty(R)$.

A "target" point $\mathbf{z} \in R^n$ is given, and we shall consider the control problem of hitting z in minimum time by trajectories of (1), using only controls $u \in U$.

Given a control u, denote the solution of (1) at time t by $\mathbf{x}(t : u)$. In fact, by variation of parameters,

$$\mathbf{x}(t : u) = X(t)\mathbf{x}_0 + X(t) \int_0^t X^{-1}(\tau)\mathbf{b}(\tau)u(\tau)\,d\tau, \tag{3}$$

where $X(t)$ is the principal matrix solution of the homogeneous system $\dot{X}(t) = A(t)X(t)$ with $X(0) = I$.

The attainable set at time t [namely, the set of all points (in R^n) reachable by solutions of (1) in time t using all admissible controls] is just

$$\mathscr{A}(t) = \{\mathbf{x}(t:u) : u \in U\} \subset R^n. \tag{4}$$

It will also be of use to consider the set

$$\mathscr{R}(t) = \left\{\int_0^t X^{-1}(\tau)\mathbf{b}(\tau)u(\tau)\,d\tau : u \in U\right\} \subset R^n. \tag{5}$$

Note that by (3)

$$\mathscr{A}(t) = X(t)[\mathbf{x}_0 + \mathscr{R}(t)] = \{X(t)\mathbf{x}_0 + X(t)\mathbf{y} : \mathbf{y} \in \mathscr{R}(t)\} \tag{6}$$

and

$$\mathscr{R}(t) = X^{-1}(t)\mathscr{A}(t) - \mathbf{x}_0, \tag{7}$$

so that $\mathbf{z} \in \mathscr{A}(t_1)$ if and only if $(X^{-1}(t_1)\mathbf{z} - \mathbf{x}_0) \in \mathscr{R}(t_1)$.

Finally, we define the bang-bang controls to be elements of

$$U_{\text{bb}} = \{u : |u(\tau)| \equiv 1 \text{ a.e. } \tau\},$$

and we denote the set of all points attainable by bang-bang controls by

$$\mathscr{A}_{\text{bb}}(t) = \{\mathbf{x}(t:u) : u \in U_{\text{bb}}\}, \qquad t > 0.$$

2. PROPERTIES OF THE ATTAINABLE SET

In this section we derive the properties of $\mathscr{A}(t)$ that will be central to the study of the control problem. Define a mapping $I : L^\infty([0, T]) \to R^n$, $T > 0$, by

$$I(u) = \int_0^T X^{-1}(\tau)\mathbf{b}(\tau)u(\tau)\,d\tau, \qquad u \in L^\infty([0, T]). \tag{1}$$

Lemma 1 I is a continuous linear mapping between $U \subset L^\infty([0, T])$ with the weak-star topology and R^n with the usual topology.

Proof The linearity follows directly from the additivity of integration.

For the continuity, recall that on bounded subsets of $L^\infty([0, T])$ the weak-star topology is metrizable, and $u_r \xrightarrow{w^*} u$ if and only if

$$\int_0^T y(\tau)u_r(\tau)\,d\tau \to \int_0^T y(\tau)u(\tau)\,d\tau \quad \text{as} \quad r \to \infty \qquad \text{for all} \quad y \in L^1([0, T]).$$

Suppose that $(u_r) \in U$ and $u_r \xrightarrow{w^*} u$. Let $X^{-1}(\tau)b(\tau) = (y_i(\tau))$. Then $\{y_i\} \in L^1([0, T])$ by assumption, and so

$$\int_0^T y_i(\tau)u_r(\tau)\,d\tau \to \int_0^T y_i(\tau)u(\tau)\,d\tau$$

for all $i = 1, 2, \ldots, n$. In other words $I(u_r) \to I(u)$. ■

As an immediate consequence of this lemma, we have the following theorem.

Theorem 1 $\mathscr{A}(T)$ is a compact, convex subset of R^n for any $T > 0$.

Proof The set of admissible controls is just the unit ball in $L^\infty([0, T])$, and so is w^*-compact and convex (Banach–Alaoglu theorem). Consequently,

$$\mathscr{R}(T) = I(u)$$

is the continuous, linear image of a w^*-compact, convex set, so must be compact and convex.

However,

$$\mathscr{A}(T) = X(T)[\mathbf{x}_0 + \mathscr{R}(T)]$$

is just an affine translate of $\mathscr{R}(T)$ in R^n, and so itself must be compact and convex. ■

A much deeper result about the structure of the attainable is contained in the following theorem, usually called the *bang-bang principle.* It says essentially that any point reachable by some admissible control in time T is reachable by a bang-bang control in the same time.

Theorem 2 (*Bang-Bang Principle*) For any $T > 0, \mathscr{A}(T) = \mathscr{A}_{\mathrm{bb}}(T)$.

Proof Note that $\mathscr{A}(T)$ is just an affine translate of $\mathscr{R}(T)$. It will then be sufficient to show that for every point $\mathbf{x} \in \mathscr{R}(T)$, $\mathbf{x} = I(u^*)$ for some bang-bang control u^*.

Let

$$B = I^{-1}(\{\mathbf{x}\}) \cap U = \{u : I(u) = \mathbf{x}\} \cap U.$$

(B is just the set of all admissible controls in U hitting $\mathbf{x}$ in time T.)

By Lemma 1, B is a weak-star compact, convex subset of $L^\infty([0, T])$, and so by Krein–Milman theorem it has an extreme point, say u^*. If we can show $|u^*| \equiv 1$ a.e., then we shall have found our bang-bang control and be finished.

Suppose not. Then there must exist a set $E \subset [0, T]$ of positive Lebesgue measure such that

$$|u^*(\tau)| < 1 \qquad \text{for} \quad \tau \in E.$$

In fact we can do a little better. Namely, let

$$E_m = \{\tau \in E : |u^*(\tau)| < 1 - 1/m\}, \qquad m = 1, 2, \ldots .$$

Then $\bigcup_{m=1}^{\infty} E_m = E$, and since E has positive measure, at least one E_m must also have positive measure. So there must exist an $\varepsilon > 0$ and a nonnull set $F \subset [0, T]$ with

$$|u^*(\tau)| < 1 - \varepsilon \qquad \text{for} \quad \tau \in F. \tag{2}$$

Since F is nonnull, the vector space $L^\infty(F)$ (again with respect to Lebesgue measure) must be infinite dimensional, and as a consequence, the integration mapping $I_F : L^\infty(F) \to R^n$,

$$I_F(v) = \int_F X^{-1}(\tau)\mathbf{b}(\tau)v(\tau)\, d\tau,$$

cannot be 1–1. (I_F maps an infinite-dimensional vector space into a finite-dimensional vector space.) Consequently, $\mathrm{Ker}(I_F) \neq \varnothing$, so we can choose a bounded measurable function $v \not\equiv 0$ on F with

$$I_F(v) = 0.$$

We now set $v = 0$ on $[0, T] - F$ so that

$$I(v) = 0, \tag{3}$$

and by dividing through by a large-enough constant we can certainly suppose $|v| \leq 1$ a.e. on $[0, T]$.

Then by (2), $|u^* \pm \varepsilon v| \leq 1$, so that $u^* \pm \varepsilon v \in U$, and by (3) $I(u^* \pm \varepsilon v) = I(u^*) \pm \varepsilon I(v) = I(u^*) = \mathbf{x}$, i.e., $u^* \pm \varepsilon v \in B$.

Since clearly,

$$u^* = \tfrac{1}{2}(u^* + \varepsilon v) + \tfrac{1}{2}(u^* - \varepsilon v)$$

and $v \not\equiv 0$, u^* cannot be an extreme point of B, a contradiction. ∎

3. EXISTENCE OF TIME OPTIMAL CONTROLS

We return to the time optimal control considered in Section 1 of this Appendix. We shall assume throughout the rest of this section that the target point $\mathbf{z}$ is hit in some time by an admissible control, that is:

($*$) There exists a $t_1 > 0$ and $u \in U$ such that $\mathbf{x}(t_1 : u) = \mathbf{z}$.

Assumptions of the type $(*)$ are called controllability assumptions, and, needless to say, are essential for the time-optimal control problem to be well posed.

If we set

$$t^* = \inf\{t : \exists u \in U \text{ with } \mathbf{x}(t : u) = \mathbf{z}\} \tag{1}$$

[this set is non empty by $(*)$ and bounded below by 0, so we can talk about its infimum], then t^* would be the natural candidate for the minimum time, and it remains to construct an optimal control $u^* \in U$ such that $\mathbf{x}(t^* : u^*) = \mathbf{z}$.

Theorem 1 If $(*)$ holds, then there exists an optimal control $u^* \in U$ such that $\mathbf{x}(t^* : u^*) = \mathbf{z}$ in minimum time t^*.

Proof Define t^* as in (1). By this definition we can choose a sequence of times $t_n \downarrow t^*$ and of controls $\{u_n\} \subset U$ such that $\mathbf{x}(t_n : u_n) = \mathbf{z}$, $n = 1, 2, \ldots$. Then

$$|\mathbf{z} - \mathbf{x}(t^* : u_n)| = |\mathbf{x}(t_n : u_n) - \mathbf{x}(t^* : u_n)|. \tag{2}$$

However, from the fundamental solution,

$$\mathbf{x}(t_n : u_n) - \mathbf{x}(t^* : u_n) = X(t_n)\mathbf{x}_0 - X(t^*)\mathbf{x}_0 + X(t_n) \int_0^{t_n} X^{-1}(\tau)\mathbf{b}(\tau)u_n(\tau)\, d\tau$$
$$- X(t^*) \int_0^{t^*} X^{-1}(\tau)\mathbf{b}(\tau)u_n(\tau)\, d\tau.$$

Consequently,

$$\begin{aligned} |\mathbf{x}(t_n : u_n) - \mathbf{x}(t^* : u_n)| \le {} & |X(t_n)\mathbf{x}_0 - X(t^*)\mathbf{x}_0| \\ & + \left| X(t_n) \int_{t^*}^{t_n} X^{-1}(\tau)\mathbf{b}(\tau)u_n(\tau)\, d\tau \right| \qquad (3) \\ & + |X(t^*) - X(t_n)| \left| \int_0^{t^*} X^{-1}(\tau)\mathbf{b}(\tau)u_n(\tau)\, d\tau \right| \end{aligned}$$

The first term on the right-hand side of (3) clearly tends to zero as $n \to \infty$. [$X(\cdot)$ is continuous.] The second term can be bounded above by

$$\|X(t_n)\| \int_{t^*}^{t_n} |X^{-1}(\tau)\mathbf{b}(\tau)|\, |u_n(\tau)|\, d\tau \le \|X(t_n)\| \int_{t^*}^{t_n} |X^{-1}(\tau)\mathbf{b}(\tau)|\, d\tau$$

since $|u_n| \le 1$. Consequently, as $n \to \infty$ this term also tends to zero. Finally, the third term tends to zero, again by the continuity of $X(\cdot)$. Plugging (3) back into (2), we get

$$\mathbf{x}(t^* : u_n) \to \mathbf{z} \qquad \text{as} \quad n \to \infty,$$

i.e., $\mathbf{z} \in \bar{\mathscr{A}}(t^*)$. However, by Theorem 2.1, $\mathscr{A}(t^*)$ is compact, and so $\mathbf{z} \in \mathscr{A}(t^*)$. Consequently,

$$\mathbf{z} = \mathbf{x}(t^* : u^*)$$

for some $u^* \in U$. ■

Note that once we have shown that an optimal control exists, the bang-bang principle guarantees that an optimal bang-bang control exists.

The above proof also carries over with only cosmetic changes to continuously moving targets $\mathbf{z}(t)$, $0 \leq t \leq T$, in fact to "continuously moving" compact target sets and to the general state equation

$$\dot{\mathbf{x}} = A(t)\mathbf{x}(t) + B(t)\mathbf{u}(t),$$

where $B(t)$ is $n \times m$ and $\mathbf{u}(t) = (u_1(t), \ldots, u_m(t))^{\mathrm{T}}$.

Appendix III Stability

Unstable control systems that magnify small errors into large fluctuations as time increases are rarely compatible with good design. Consequently, tests for the stability of systems of differential equations are extremely important.

The system of differential equations

$$\dot{\mathbf{x}} = \mathbf{f}(\mathbf{x}) \tag{1}$$

is said to be *stable* about $\mathbf{x} = \mathbf{0}$ if, for every $\varepsilon > 0$, there exists a $\delta > 0$ such that $\|\mathbf{x}_0\| < \delta$ implies that the solution $\mathbf{x}(t)$ starting at $\mathbf{x}(0) = \mathbf{x}_0$ obeys $\|\mathbf{x}(t)\| < \varepsilon$ for $0 \leq t < \infty$.

A necessary condition for stability about $\mathbf{0}$ is clearly that $\mathbf{f}(0) = \mathbf{0}$, or that $\mathbf{0}$ is an *equilibrium* point of (1). Usually, for control applications, one requires more than just stability; one would prefer that small fluctuations be damped out as time increases.

The system (1) is called *asymptotically stable* about the origin if, for every $\varepsilon > 0$, there exists a $\delta > 0$ such that $\|\mathbf{x}_0\| < \delta$ implies that $\mathbf{x}(t) \to \mathbf{0}$ as $t \to \infty$, where $\mathbf{x}(t)$ is the solution of (1) starting at $\mathbf{x}(0) = \mathbf{x}_0$. Furthermore, (1) is *globally asymptotically stable* about $\mathbf{0}$ if every solution of (1) converges to $\mathbf{0}$ as $t \to \infty$. Global asymptotic stability is the most useful concept for practical design of control systems. If (1) is linear, that is, if

$$\dot{\mathbf{x}} = A\mathbf{x}, \tag{2}$$

it is easy to see that a necessary and sufficient condition for global asymptotic stability is that A have eigenvalues with negative real parts only. For if $\{\lambda_j\}_{j=1}^P$ are the distinct eigenvalues of A, then any solution of (2) can be written

$$\mathbf{x}(t) = \sum_{j=1}^{P} \mathbf{g}_j(t) \exp(\lambda_j t), \tag{3}$$

where the $\mathbf{g}_j(t)$ are polynomials in t. From (3) it is easy to see that $\mathbf{x}(t) \to 0$ if and only if $\operatorname{Re} \lambda_j < 0, j = 1, 2, \ldots, P$.

The eigenvalues of A are determined as the roots of the characteristic equation $\det(\lambda I - A) = 0$. Once this equation has been computed it is not necessary to find all of the roots; the famous Routh–Hurwitz test gives necessary and sufficient conditions for the roots of a polynomial

$$d(\lambda) = a_0\lambda^n + a_1\lambda^{n-1} + \cdots + a_{n-1}\lambda + a_n, \qquad a_0 > 0, \tag{4}$$

to have negative real parts (for a proof, see [1, Vol. II, Chapter XV]. First, define the following determinants

$$\Delta_k = \begin{vmatrix} a_1 & a_3 & a_5 & & & \\ a_0 & a_2 & a_4 & & & \\ 0 & a_1 & a_3 & & & \\ 0 & a_0 & a_2 & a_4 & & \\ 0 & 0 & a_1 & a_3 & & \vdots \\ 0 & 0 & 0 & a_2 & \ddots & \\ \vdots & \vdots & \vdots & \vdots & & \\ 0 & 0 & 0 & 0 & & a_k \end{vmatrix}, \qquad k = 1, 2, \ldots, n,$$

where we substitute zero for a_k if $k > n$.

Theorem 1 The polynomial (4) with real coefficients has all its roots with negative real parts if and only if the coefficients of (4) are positive and the determinants $\Delta_k > 0$, $k = 1, 2, \ldots, n$.

Example 1 For the characteristic equation

$$a_0\lambda^4 + a_1\lambda^3 + a_2\lambda^2 + a_3\lambda + a_4 = 0,$$

the conditions for asymptotic stability are

$$a_i > 0, \qquad i = 0, 1, 2, 3, 4,$$

and

$$\Delta_2 = \begin{vmatrix} a_1 & a_3 \\ a_0 & a_2 \end{vmatrix} = a_1 a_2 - a_0 a_3 > 0,$$

$$\Delta_3 = \begin{vmatrix} a_1 & a_3 & 0 \\ a_0 & a_2 & a_4 \\ 0 & a_1 & a_3 \end{vmatrix} = a_3(a_1 a_2 - a_0 a_3) - a_1^2 a_4 > 0.$$

Note that since $a_i > 0$, $\Delta_2 > 0$ follows from $\Delta_3 > 0$.

Index

N

O

P

R

S

T

V

W